New Insights, Trends, and Challenges in the Development and Applications of Microbial Inoculants in Agriculture

Developments in Applied Microbiology and Biotechnology

New Insights, Trends, and Challenges in the Development and Applications of Microbial Inoculants in Agriculture

Edited by

Sergio de los Santos Villalobos
Instituto Tecnologico de Sonora, Ciudad Obregon, Sonora, Mexico

Academic Press is an imprint of Elsevier
125 London Wall, London EC2Y 5AS, United Kingdom
525 B Street, Suite 1650, San Diego, CA 92101, United States
50 Hampshire Street, 5th Floor, Cambridge, MA 02139, United States

Notices
Knowledge and best practice in this field are constantly changing. As new research and experience broaden our understanding, changes in research methods, professional practices, or medical treatment may become necessary.

Practitioners and researchers must always rely on their own experience and knowledge in evaluating and using any information, methods, compounds, or experiments described herein. In using such information or methods they should be mindful of their own safety and the safety of others, including parties for whom they have a professional responsibility.

To the fullest extent of the law, neither the Publisher nor the authors, contributors, or editors, assume any liability for any injury and/or damage to persons or property as a matter of products liability, negligence or otherwise, or from any use or operation of any methods, products, instructions, or ideas contained in the material herein.

ISBN: 978-0-443-18855-8

For information on all Academic Press publications visit our website at https://www.elsevier.com/books-and-journals

Publisher: Stacy Masucci
Acquisitions Editor: Kattie Washington
Editorial Project Manager: Billie Jean Fernandez
Production Project Manager: Gomathi Sugumar
Cover Designer: Greg Harris

Typeset by TNQ Technologies

Contents

Part III
Present and future scenario

Contributors

Alondra María Díaz-Rodríguez, Instituto Tecnológico de Sonora, Ciudad Obregón, Sonora, Mexico

Amelia C. Montoya-Martínez, Instituto Tecnológico de Sonora, Ciudad Obregón, Sonora, Mexico

Valeria Valenzuela Ruiz, Instituto Tecnológico de Sonora, Ciudad Obregón, Sonora, Mexico

Marisol Ayala-Zepeda, Instituto Tecnológico de Sonora, Ciudad Obregón, Sonora, Mexico

Jonathan Rojas-Padilla, Instituto Tecnológico de Sonora, Ciudad Obregón, Sonora, Mexico

Luis Abraham Chaparro-Encinas, Universidad Autónoma Agraria Antonio Narro, Torreón, Coahuila, Mexico

Roel Alejandro Chávez-Luzanía, Instituto Tecnológico de Sonora, Ciudad Obregón, Sonora, Mexico

Eber D. Villa-Rodríguez, Aarhus University, Aarhus, Denmark

María Fernanda Villarreal-Delgado, Sartorius de México, Azcapotzalco, Ciudad de México, Mexico

Fannie Isela Parra Cota, Campo Experimental Norman E. Borlaug, INIFAP, Ciudad Obregón, Sonora, Mexico

Alejandra Miranda Carrazco, Departamento de Ciencias Ambientales, UAM Unidad Lerma (UAML), Lerma Estado de México, Mexico

Lorena Jacqueline Gómez-Godínez, Centro Nacional de Recursos Genéticos, INIFAP, Tepatitlán de Morelos, Jalisco, Mexico

Edith Rojas Anaya, Centro Nacional de Recursos Genéticos, INIFAP, Tepatitlán de Morelos, Jalisco, Mexico

Ramón Arteaga Garibay, Centro Nacional de Recursos Genéticos, INIFAP, Tepatitlán de Morelos, Jalisco, Mexico

José Martín Ruvalcaba Gómez, Centro Nacional de Recursos Genéticos, INIFAP, Tepatitlán de Morelos, Jalisco, Mexico

Hugo Alberto Zaldivar-López, Centro Nacional de Recursos Genéticos, INIFAP, Tepatitlán de Morelos, Jalisco, Mexico

Sergio de los Santos Villalobos, Instituto Tecnológico de Sonora, Ciudad Obregón, Sonora, Mexico

Preface

In the last three decades, a growing interest in having agroecological alternatives to stimulate the growth, yield, quality, and health of crops has been widely documented. One of them is the use of microbial inoculants, which, used properly, can provide great benefits to the crops where they are applied, but without the toxic and polluting effects of agrochemicals (pesticides or synthetic fertilizers). In addition, microbial inoculants, which mostly consist of plant growth-promoting bacteria or fungi, can exert mechanisms to ameliorate different types of environmental stress, such as salinity, drought, heat, or heavy metal contamination, to name a few.

In this book, different works by leading scientists in their field have been brought together to write 15 chapters that range from basic aspects of plant–microorganism interactions to different options for bioprospecting and applying microbial inoculants in different agroecosystems. But not everything in the research on these topics is an easy task; currently, there are still several challenges that must be overcome to have wide use of beneficial microorganisms in the agricultural fields worldwide. Thus, this book analyzes these vicissitudes and proposes strategies for the extensive development and application of next-generation microbial inoculants for sustainable agriculture.

The chapters of this book are written pleasantly, with well-explained concepts, and excellent figures, which are highly didactic. In this sense, I do not doubt that this work will become a reference for researchers, academics, undergraduate and graduate students, as well as stakeholders and policymakers interested in delving into the world of microbial inoculants. Here, the interested community will find a series of microbial, biochemical, molecular, and ecological knowledge, as well as new omic technologies that are novel and widely used in recent literature.

I congratulate and thank the Editor, Dr. Sergio de los Santos Villalobos, for the invitation to write these lines (as well as the scoop to enjoy reading) and for the initiative to integrate all this knowledge in a single work, as well as the authors, whose effort has been consolidated in excellent work.

Gustavo Santoyo Pizano
Universidad Michoacana de San Nicolas de Hidalgo, Mexico

Acknowledgments

Deep gratitude is extended to all the collaborators of this book for their valuable work, which contribute to the advance of the knowledge for the bioformulation of sustainable and biosafety next-generation microbial inoculants to food security and safety worldwide. Furthermore, recognition is given to the Technological Institute of Sonora (ITSON), as well as all participating institutions for their support and endorsement of the academic–scientific work of those involved in this project, such as the Antonio Narro Agrarian Autonomous University (UAAAN), Aarhus University, Sartorius-Mexico, National Institute for Forest, Agriculture and Livestock Research (INIFAP), and National Genetic Resources Center (CNRG). Lastly, we thank the National Council of Humanities, Science, and Technology (CONAHCYT); the Regional Agreement for the Promotion of Nuclear Science and Technology in Latin America and the Caribbean (ARCAL), sponsored by the International Atomic Energy Agency (IAEA); the General Directorate of Soils and Water belonging to the Secretariat of Agriculture and Rural Development (SADER); and the Agricultural Research and Experimentation Board of the State of Sonora A.C. (PIEAES A.C.) for funding and training provided to consolidate and advance the use of cutting-edge strategies aimed at contributing to the global sustainable food security. This book is dedicated to the generation of knowledge, technologies, and comprehensive training of highly critical and leadership-oriented human resources to face the current and future challenges of our society.

Sergio de los Santos Villalobos
Editor

Chapter 1

Introduction

Alondra María Díaz-Rodríguez[1], Edith Rojas Anaya[2], Marisol Ayala-Zepeda[1] and Sergio de los Santos Villalobos[1]

[1]*Instituto Tecnológico de Sonora, Ciudad Obregón, Sonora, Mexico;* [2]*Centro Nacional de Recursos Genéticos, INIFAP, Tepatitlán de Morelos, Jalisco, Mexico*

Agricultural systems cover nearly 5 billion hectares of the global terrestrial land area, but the accelerated growth of the human population has forced farmers to increase food production in the same area (Ritchie & Roser, 2019). In 2020, the world population exceeded 7.7 billion inhabitants, and it is estimated that the world population will reach close to 10 billion by 2050, therefore food production needs to increase by 70%−100% by that year (FAO, 2017).

To meet the needs of the current population, farmers have implemented intensive production systems proposed during the Green Revolution; for example, the use of improved seed varieties resistant to diseases and optimized agricultural management practices such as planting densities, tillage methods, irrigation schedules, chemical fertilization, and phytosanitary control by the use of pesticides, among others (Matson, 2012). Thus, between 1960 and 2000, an increase in yield per hectare was obtained rather than an expansion of area under cultivation for all developing countries, for example, 208% for wheat, 109% for rice, 157% for maize, and 78% for potatoes (FAO, 2004). However, the long-term yield of some crops has been decreasing in the last years compared to the productivity obtained during the Green Revolution, indicating that current intensive practices are not sustainable enough to meet the future global food demand. From the year 2000 to 2018, yields only reached 37% for maize, 29.1% for potatoes, 25.75% for wheat, and 20.36% for rice (Ritchie & Roser, 2019).

Moreover, these conventional practices imply the consumption of nonrenewable resources, the use of chemical fertilizers and pesticides, and the runoff or leaching of soluble nutrients from the agrochemicals into the aquatic systems, which among other consequences, are sources of environmental contamination, represent a high economic investment by the producers, and present risks to human health (Fig. 1.1) (Barea, 2015; Matson, 2012).

New Insights, Trends, and Challenges in the Development and Applications of Microbial Inoculants in Agriculture
https://doi.org/10.1016/B978-0-443-18855-8.00001-1

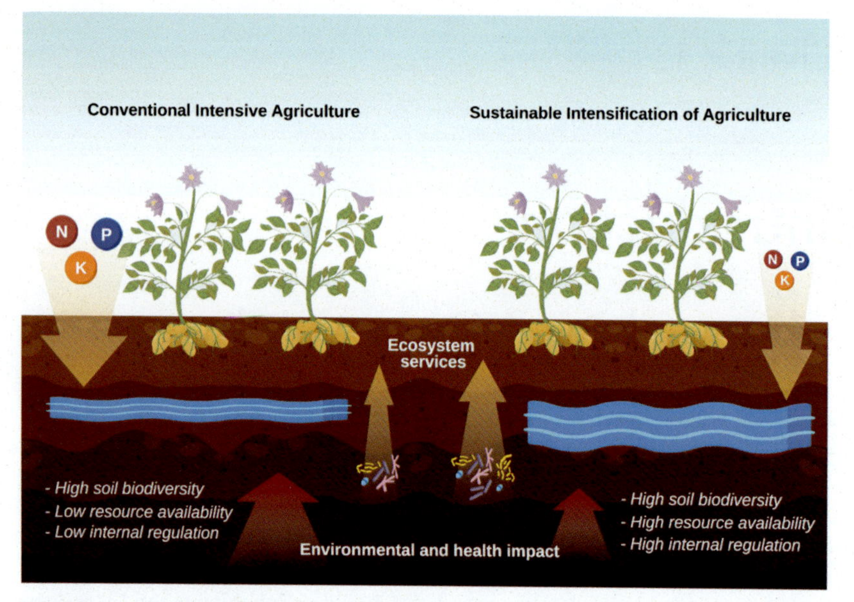

FIGURE 1.1 Conceptualization of a sustainable cropping system combining features of intensive cropping systems to meet the challenge to improve yields and quality of the crops.

In addition, according to the Intergovernmental Panel on Climate Change (IPCC, 2013), the effect of climate change will cause an increase in the average temperature of around 1.4–5.8°C and modifications in the intensity and temporal distribution of precipitations during the current century (2001–2100) (Cubasch & Wuebbles, 2013). These predictions could reduce crop yield due to changes in the soil ecology (including plant–microbe interactions), changes in plant growth, transpiration, respiration, and photosynthesis rates, and alterations in the distribution and proliferation regimes of new phytopathogens, pests, and weeds (Mall et al., 2017).

Therefore, climate change forces farmers to rely more on agrochemicals to maintain yields and protect crops from diseases and pests, becoming a vicious cycle since the alteration in agricultural land and intensive practices decrease soil productivity. This consequently generates (1) high economic costs for farmers, for example, synthetic fertilizers represent 30%–40% of the total cost for wheat and maize production (USDA, 2022); and (2) environmental risks due to leaching, volatilization, generation of greenhouse gases, soil salinity, contamination of groundwater, and nutritional and ecological imbalance of the soil, mainly affecting the native microbiota (Zhang et al., 2018).

Nowadays, economic and social challenges are faced in reducing climate change and environmental impacts that directly affect food production (Hodson de Jaramillo, 2018). Therefore, the generation and application of sustainable agricultural strategies adapted to the predicted climate change

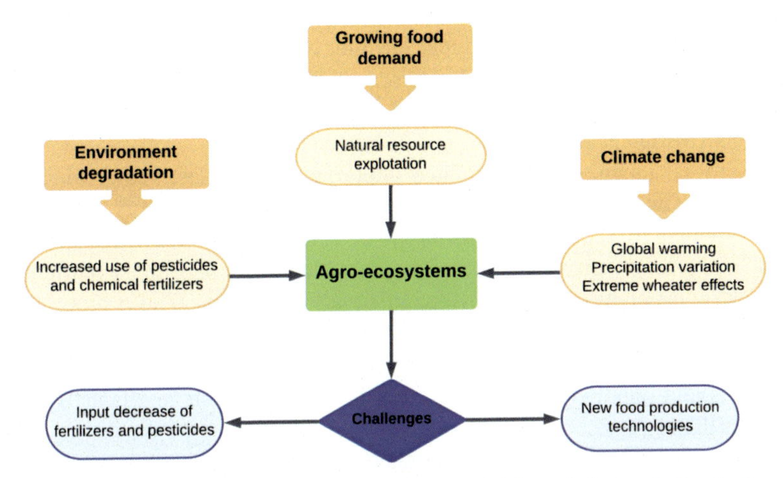

FIGURE 1.2 Relation between agriculture, environmental degradation, and climate change. Within the majority of current agricultural systems, intensive agriculture is carried out, mainly led by the continuous entry of high amounts of pesticides and nitrogen fertilizers, which increases the extension of degraded ecosystems.

scenarios that decrease the use of synthetic fertilizers and pesticides, increase crop yields, and contribute to actual and future food security are needed (Moreno-Reséndez et al., 2018) (Fig. 1.2).

Several authors have reported that intensive agricultural practices affect the functions of agroecosystems, and these are enhanced by the effects of climate change, such as increases in temperature, changes in precipitation patterns, and extreme weather events (e.g., frost and droughts) that result in lower yields. Together, adding to the growing demand for food, they create strong pressure on natural resources (e.g., soils, water, flora, fauna, and edaphic microbial resources), which has a particularly negative impact on agricultural systems and goods and services they provide us. Therefore, the current challenge is to adopt more efficient production methods that can adapt to climate change and be more productive to contribute to food security. One strategy is sustainable intensification (SI), including new agro-biotechnological approaches (Fig. 1.1).

SI principles include enhancing resource use efficiency, conservation and restoration of agroecosystems, diversifying production systems, knowledge and technology transfer, and socio-economic considerations. Therefore, these agro-biotechnological approaches must consider the collaboration between experts and stakeholders, where the knowledge and needs of farmers should be prioritized. Besides, to promote easy access to appropriate information, training, and innovative technologies through conventional extension (research to farmers) and participatory dissemination (Pretty & Bharucha, 2014). In addition, the establishment of political will is crucial for the regulation and promotion of sustainable production alternatives, which can be explored within the framework of the bio-economy

(Hodson de Jaramillo, 2018). Some new approaches that can be considered include developing more ecological farming practices, exploring the genetic variability in our existing food crops, and studying the genetic and metabolic potential of the soil microbiota for the production of microbial inoculants (Mitter et al., 2019).

Microbial inoculants have gained importance in the last years, which are eco-friendly and sustainable bioproducts containing live microorganisms that, when applied to seeds, plant surfaces, or soil, colonize the rhizosphere or plant, and promote growth by increasing the availability of primary nutrients to the host plant or by protecting it against phytopathogens. These bioproducts are biologically active and contain one or more beneficial microorganisms in an easily applicable and economical carrier material, maintaining cell viability and sufficient population to exert growth-promoting effects on plants (Malusá et al., 2012).

Soil microbial communities can interact with crops, regulating their growth and productivity, through their tolerance to abiotic and biotic stress, plant nutrition, and antagonism of plant pathogens. These microorganisms are commonly known as plant growth promotion microorganisms (PGPM) (Santoyo et al., 2019). PGPM are a heterogeneous group of microorganisms and include plant growth-promoting fungi (PGPF), arbuscular mycorrhizal fungi (AMF), biological control agents (BCA), and plant growth-promoting bacteria (PGPB). Over the years, an increasing number of PGPM are being used as components of sustainable agricultural practices. Microorganisms have been extensively isolated from soil and host plants to test their capabilities related to plant growth (biofertilizers) and defense against pathogens (biocontrol agents) to prospect bacterial inoculants (Ambrosini et al., 2016).

Microbial inoculants may be a partial or complete substitute for chemical fertilization and pesticides, offering an environmentally sustainable approach to increasing crop production and overall eco-systemic health (plant health, growth, and soil fertility) (Gupta et al., 2015). These bioproducts have been developed based on the ability of PGPM to interact with the plants and their microbiomes and can be classified based on their studied functions, for example, nitrogen-fixing bacteria, phosphate solubilizers, micronutrients biofertilizers, biopesticides, among others. Besides, they can be formulated using a single microbial strain that exhibits the desired function, or developed as a combination of different strains or microbial consortia. The choice of formulation depends on various factors, including the target crop, specific edaphoclimatic conditions, and desired effects (Timmusk et al., 2017).

Microbial inoculants have the potential to increase crop yield by up to ~ 10%–30% and reduce the use of NPK (nitrogen, phosphorus, and potassium) fertilizers by up to 50% along with yield increase (Nadeem et al., 2014; Nguyen et al., 2017; Thilagar et al., 2016). Successful experiments have been conducted in vitro, in greenhouses, and under field conditions, applying inoculants based on PGPM such as *Bacillus*, *Pseudomonas*, *Burkholderia*,

Bradyrhizobium, *Trichoderma*, *Aspergillus*, *Funneliformis*, and *Rhizoglom*us (Dong et al., 2019; Lu et al., 2020; Rojas-Padilla et al., 2020; Thilagar et al., 2016); however, their results sometimes vary due to the complex field environment where various factors act simultaneously (Timmusk et al., 2017).

Therefore, the exploration of native edaphic and/or endophytic bacteria for the formulation of microbial inoculants represents a promising sustainable alternative to address food security, which has been the main reason for the failure of the imported bioproducts. In this sense, it is important to consider that the formulations of inoculants should integrate some characteristics: provide a suitable microenvironment for growth, ensure the viability of the microbial components, ensure the competence of the inoculant in the microflora of the native soil, and, if possible, avoid losses caused by microfauna. The above ensures an adequate interaction between plants and the soil microbiome (Bashan et al., 2014).

This book describes the role of microbial inoculants (specifically PGPB and BCA) as a strategy to promote sustainable agriculture, emphasizing the key points for the development and formulation of high-quality bacterial inoculants, as well as several ecological and functional considerations for their application in the field. Finally, the importance of a legal framework to guarantee the quality and safety of these bioproducts is discussed, as well as participatory innovation frameworks from all sectors, mainly agricultural producers, for the success of this alternative.

Part I

Microbial ecology in food security

Chapter 2

Microbial interactions in agroecosystems

Lorena Jacqueline Gómez-Godínez[1], Alondra María Díaz-Rodríguez[2], Fannie Isela Parra Cota[3] and Sergio de los Santos Villalobos[2]
[1]*Centro Nacional de Recursos Genéticos, INIFAP, Tepatitlán de Morelos, Jalisco, Mexico;* [2]*Instituto Tecnológico de Sonora, Ciudad Obregón, Sonora, Mexico;* [3]*Campo Experimental Norman E. Borlaug, INIFAP, Ciudad Obregón, Sonora, Mexico*

Soil and plants host a large diversity of microorganisms, known as microbiota, such as bacteria, fungi, actinomycetes, algae, protozoa, and viruses, which can be either pathogenic or beneficial. The total biomass composition of the biosphere is estimated to be $\approx$550 gigatons of carbon (Gt C); of these, plants have a biomass of approximately $\approx$450 Gt C, followed by bacteria with $\approx$70 Gt C (Bar-On et al., 2018). The soil is considered one of the most complex and diverse ecological systems existing on the planet (Islam et al., 2020). It is considered that the richness of species and the diversity of the soil microbiome are very high, estimating that 1 g of soil may harbor up to 10^9 microbial cells (Bhattarai, 2015), belonging to 10^4-10^6 different species (Roesch et al., 2007). This soil microbiota is a key component in maintaining soil fertility and provides socioeconomic and ecological benefits, for example, climate change mitigation, provision of biotechnological resources, and food production (Díaz-Rodríguez et al., 2021).

A better understanding of the soil–plant–microbe interactions and the beneficial functions of plant microbiota could potentially lead to the development of microbiome-based solutions such as microbial inoculants (Mitter et al., 2019). An important component of the plant microbiota includes plant growth-promoting bacteria (PGPB) that reside in the rhizosphere, phyllosphere, and as endophytic bacteria (Compant et al., 2019).

The rhizosphere is the soil portion that is influenced by root exudates. Plants commonly exude from $\sim$11% to $\sim$40% of all the carbon fixed by photosynthesis, including organic acids, amino acids, sugars, hormones, enzymes, and products of secondary metabolism, among others, through their roots leading to a high availability of nutrients for a large range of

New Insights, Trends, and Challenges in the Development and Applications of Microbial Inoculants in Agriculture
https://doi.org/10.1016/B978-0-443-18855-8.00002-3

microorganisms (Sun et al., 2021). For this reason, the concentration of microbes around the roots is approximately 10 to 1000-fold greater than that found in the bulk soil (Glick, 2020).

The importance of the microbial interactions in the rhizosphere lies in their contribution to ecosystem functions (Fig. 2.1), such as their role in organic matter decomposition, nutrient cycling (especially for nutrients that are hardly available to plants, such as nitrogen and phosphorus), vegetal biomass production, water cycling, and soil aggregation and fertility (Neemisha, 2020). The plant microbiome structure is influenced by interactions between plant hosts (e.g., genotype, developmental stage, and root exudates), microorganisms (e.g., metabolites and signaling), environmental factors, and cultivation practices (Dastogeer et al., 2020). Various reports affirm that intensive agricultural practices are directly related to groundwater contamination, air pollution, and a decrease in the diversity of beneficial microorganisms (Gómez-Godínez et al., 2021). For example, Lares-Orozco et al. (2016) found that fertilization was the main source of greenhouse gas (GHG) emissions (83%) in the life cycle of wheat in the Yaqui Valley, Mexico, which is characterized by intensive crop production. Also, according to Shen et al. (2016), the use of a conventional nitrogen fertilization rate over a decade significantly reduced the richness of bacterial Operational Taxonomic Unit (OTU) phylotypes and the phylogenetic diversity in soils situated in China's Yangtze River Delta.

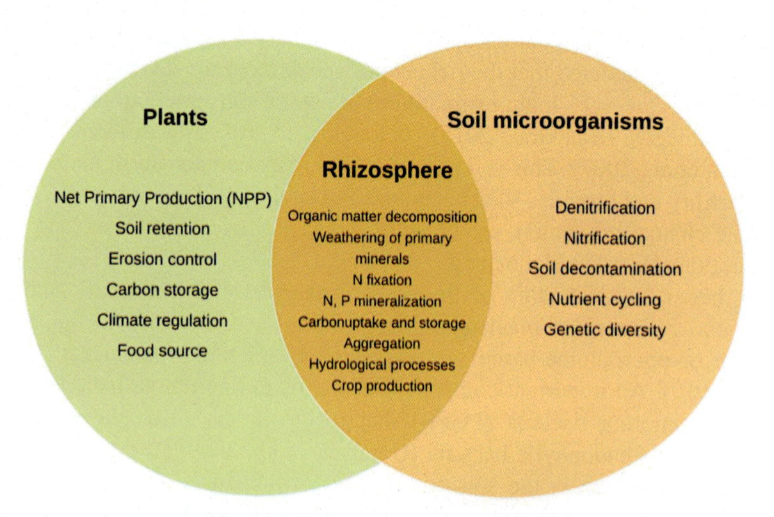

FIGURE 2.1 Ecosystem services and functions in the rhizosphere due to plant–microbial interactions. Rhizosphere processes contribute to a variety of ecosystem services and are the outcome of plant–microbial interactions. Processes such as aggregation, nutrient cycling, hydrology, and carbon storage are jointly mediated by plants and soil microorganisms through interactions in the rhizosphere.

PGPB have been shown to promote plant growth, inhibit pathogens, and modify the structure of soil microbial communities, which has made them very popular commercially (Ambrosini et al., 2016). These microorganisms have been successful in various applications in the colonization of soil, roots, and plants. Compared to conventional fertilizers and pesticides, PGPB-based bioformulations can improve plant health and growth and, in turn, promote the abundance of potentially beneficial microbial taxa and increase enzymatic activities in the soil, such as alkaline phosphatase and invertase activity, which are involved in the process of carbon sequestration and organic phosphorus mineralization (Dong et al., 2019; Lu et al., 2020).

Another activity that is favored is the action of enzymes such as nitrogenase, which has different structural subunits (molybdenum, iron, and vanadium), responsible for transforming atmospheric nitrogen into assimilable nitrogen (Aasfar et al., 2021). There are other low molecular weight molecules, known as siderophores. These molecules have a high affinity to bind iron (Fe III) or other metals, making them available to plants or other microorganisms (Ahmed & Holmström, 2014). Phosphatases are also released into the soil; these are enzymes responsible for the decomposition and mineralization of phosphorus by catalyzing the hydrolysis of esters and anhydrides of phosphoric acid. They are generally classified as phosphomonoesterase, phosphodiesterases, and enzymes that act on anhydrides containing phosphoryl or on phosphorus–nitrogen bonds. They are responsible for making phosphorus bioavailable in the soil and these are produced by different microorganisms (Tian et al., 2021). Other factors such as soil pH, edaphic conditions, plant genotype, and native microbiota are also related to the performance of microbial inoculants (Gómez-Godínez et al., 2021).

The phyllosphere comprises the aerial parts of plants and is dominated by the leaves. It has been estimated that there is an average of 10^6–10^7 bacteria per square cm of the leaf surface, and that the global bacterial population present in the phyllosphere could reach up to 10^{26} (Vorholt, 2012). Microbial communities found in the phyllosphere play important roles related to plant development by affecting the functions and longevity of leaves, seed mass, apical growth, flowering, and fruit development. They also play key roles in contaminant removal, stress resistance, and phytohormone biosynthesis. For example, bacteria of the genus *Microbacterium*, *Stenotrophomonas,* and *Methylobacterium* can improve the growth and nutritional status of the host plant through the production of natural growth regulators such as indole acetic acid, and nitrogen fixation (Abadi et al., 2020). The phyllosphere microbiome is involved in maintaining plant health and suppressing plant pathogen overgrowth or disease suppression. It also plays an important role in reducing emissions of vegetable methanol and isoprene to the atmosphere (Fig. 2.3) (Zhu et al., 2022). Bacteria and yeasts present in the phyllosphere, capable of colonizing nectar, can modulate its chemical composition and, consequently, influence the attraction of pollinating insects (Liu et al., 2020).

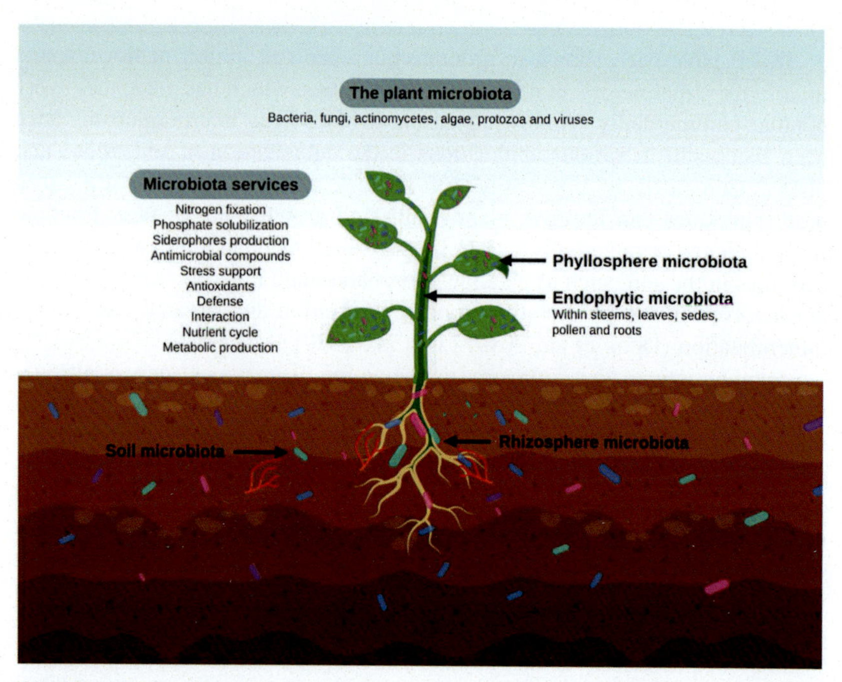

FIGURE 2.2 The microbiota present in the soil, the rhizosphere, the phyllosphere, and as endophytes provide different services and benefits to the plant and to the environment where it operates.

Also, microorganisms living within the plant, including aboveground and underground parts and seeds, known as endophytes, perform essential functions for plant growth and development by the modulation of metabolic interactions, nutrient uptake, enhancing stress tolerance and adaptation (Fig. 2.2); (Parasuraman et al., 2019). Root endophytes possess mechanisms that include: (1) production of phytohormones, (2) biological control of phytopathogens, (3) supply of nitrogen or phosphate nutrients for plants, (4) siderophore production, (5) nitrogen fixation, (6) phosphate solubilization, and (7) reduction of ethylene stress by synthesizing 1-aminocyclopropane-1-carboxylate (ACC) deaminase (Vandana et al., 2021). Seeds are privileged carriers of beneficial bacteria for plants and are even identified as critical points for the isolation of beneficial bacteria, as they show the ability to produce phytohormones such as indole acetic acid, gibberellins, and cytokines, either individually or in combination. These bacteria enable increased nutrient assimilation by plants and enhance their resistance to salt stress, drought, heat, frost, heavy metals, and pests. Seed endophytic bacteria play a key role in seedling germination and growth, influencing seed viability, germination, and seedling survival (L'Hoir & Duponnois, 2021).

The soil–plant–microbe interactions have been shown to have beneficial effects on plants, including increased nutrient acquisition (Valenzuela-

Aragon et al., 2019), disease suppression (Villa-Rodriguez et al., 2021), induction of systemic resistance (Fatima & Anjum, 2017), and tolerance to abiotic stress (Egamberdieva et al., 2017), leading to an increase in yield. Therefore, the study and understanding of the molecular, metabolic, and genetic bases of plant–microorganism interactions serve as a tool to develop truly effective and long-lasting solutions for the current and future food demand, as well as the need for better and more sustainable crop production. The use and integration of omics techniques have allowed a close molecular insight and a better comprehension of the relationships established between plants and microorganisms. In this way, the study and use of PGPB and biological control agents (BCA) are promising strategies to enhance crops' health and growth by contributing to nutrient supply and resilience under environmental (biotic and abiotic) stress conditions (Mitter et al., 2019).

Through time, and thanks to omics sciences, significant contributions have been made to the understanding of the structure, composition, abundance, and importance of microbiomes and their interaction in agroecosystems. For instance, the diversity of the microbiome in the rhizosphere of different maize genotypes and different types of soil has been reported, finding that microbial communities change depending on the plant genotype, root exudates, and soil types. In addition, there is a tendency for the soil microbiome to decrease when it is subjected to adverse conditions (Peiffer et al., 2013; Philippot et al., 2013). Metabarcoding approaches have revealed variations in the bacterial and fungal communities depending on the agricultural management of the soil, such as horticultural, agricultural, and vineyard soils. Vineyards revealed a significantly higher diversity and distinct composition of soil fungi compared to horticultural and agricultural soils. The abundance of the genus *Solicoccozyma*, a yeast commonly found in soils with high salt content, was also reported in horticultural soils, most likely due to fertilization management practices, which generally involve higher application rates in orchards (Köberl et al., 2020).

We can explore microbial communities and their metabolism when they are subjected to different types of stress. Xu et al. (2018) reported the ecological succession of the bacterial community in the roots of sorghum (*Sorghum bicolor*) under drought conditions, and they observed that drought delayed the normal development of the root microbiome and the rhizosphere. Drought stress led to a strong enrichment of gram-positive bacteria, including Actinobacteria and Firmicutes, and microorganisms of the phylum Chloroflexi. Under these conditions, the overexpression of genes related to transport and catabolism of carbohydrates, amino acids, and secondary metabolites, many of which are known to be present in root exudates produced by rhizosphere-feeding plants and root-associated microorganisms, suggests that drought-induced changes in root metabolism are related to changes in the composition of the root-associated microbiome (Xu et al., 2018).

Some treatments are focused on soil amendment, such as wood ash, which is known to have beneficial effects as it counteracts acidification and returns essential nutrients to the soil. In addition, it also generates changes in the microbial composition, favoring copiotrophic groups such as Bacteroidetes, Alphaproteobacteria, and Betaproteobacteria, as these are initial metabolizers of labile carbon and respond positively to the increase in soil pH and electrical conductivity. Before any change in the environment, the microorganisms respond, and in this case, the genes are overexpressed in response to stress, sporulation, and carrier proteins that help balance the osmotic pressure of cells, regulate cytosolic pH, and can export toxins such as metals from the cell (Bang-Andreasen et al., 2019).

Synthetic communities of more than two microorganisms have also been created to evaluate their effects on plants and the expression of genes during the interactions between members of the synthetic community and the plant. For example, Gómez-Godínez et al. (2019), in an evaluation of five bacteria with the potential to promote growth, in maize plants, found overexpressed genes related to biological nitrogen fixation, biofilm production, metabolite transport, and siderophore production. These gene overexpressions were related to the growth of maize plants and the interactions between microorganisms present in the synthetic community (Gómez-Godínez et al., 2019).

Exometabolomics is an approach to determining the metabolites produced or depleted in a given environment (Allen et al., 2003). This approach allows us to delve into the plant–microorganism and microorganism–microorganism interactions. Exometabolomics has provided insights into the interaction between *Pseudomonas syringae* and *Arabidopsis thaliana*, finding that *Pseudomonas* infection can reprogram plant signaling and primary metabolic pathways involved in stomatal movement. Furthermore, it increases carbohydrates and ATP and decreases the presence of amino acids; fructose showed a more than threefold increase in this interaction, and most amino acids showed a decrease after 180 minutes of interaction with *Pseudomonas*. These findings suggest that carbohydrate metabolism responds rapidly to the plant–bacteria interaction, while changes in amino acid levels occur later (Pang et al., 2018).

Zhao et al. (2022) evaluated the metabolites of the rhizosphere microbiome associated with barley crops over different periods, specifically short-term (2 years) and long-term (10 years), finding changes in the metabolites, mainly in lipids and lipid-like molecules, acids organic and derivatives, and organoheterocyclic compounds. Among these, lipids and lipid-like molecules had the highest amount of metabolites affected by long-term cultivation. Furthermore, nucleosides, nucleotides, and analogs were mostly upregulated over time, while alkaloids and derivatives were mostly downregulated. All these metabolites were directly correlated with the different orders and families present in the soil associated with the crop, finding that this production of metabolites is mainly due to the orders Desulfuromonadales and Nitrosomonadales, the

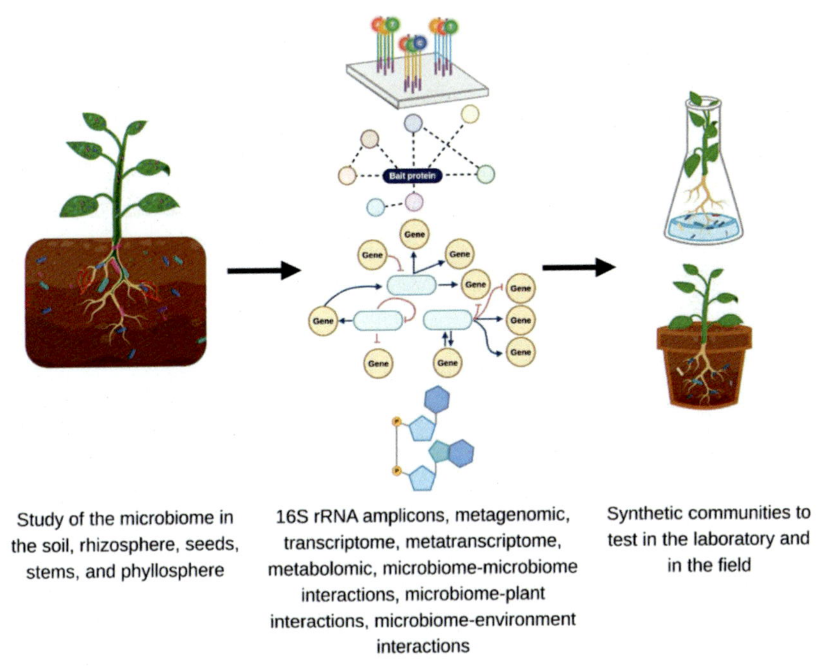

FIGURE 2.3 An integrated perspective on the function and interaction of the microbiome, which combines omics strategies, such as amplicon sequencing, metagenomics, transcriptomics, metatranscriptomics, and metabolomics, to improve the designs of synthetic communities, which can have synergistic effects on agroecosystems.

families Archangiaceae, and Nocardiacea, among others. This study suggests that soil metabolomics can be used to assess the adaptations of soil microbial communities to short- and long-term cultivation strategies at a molecular level.

The interaction between the microbiome and its environment is dynamic; they are always interacting with each other and with the surrounding environment. There is a need for integrative studies between microbiomes and environments, considering the different mechanisms involved. Such studies will aid in the management of the microbiota and the prediction of interactions, ultimately leading to improving the sustainability of food production (Fig. 2.3).

Chapter 3

Plant–pathogen interactions: Mechanisms involved in plant diseases

Amelia C. Montoya-Martínez[1], Valeria Valenzuela Ruiz[1], Roel Alejandro Chávez-Luzanía[1], Fannie Isela Parra Cota[2] and Sergio de los Santos Villalobos[1]

[1]*Instituto Tecnológico de Sonora, Ciudad Obregón, Sonora, Mexico;* [2]*Campo Experimental Norman E. Borlaug, INIFAP, Ciudad Obregón, Sonora, Mexico*

Plant diseases limit the potential crop production and cause considerable losses in agriculture, horticulture, and forestry. Plants are healthy when they perform their physiological functions at their maximum genetic potential, functions such as cell division, differentiation, and development, absorption of water and minerals, photosynthesis, metabolism of synthesized compounds, reproduction, and storage of the food reserves necessary for reproduction or wintering (Agrios, 1995). Deviations from these normal functions can occur, and this abnormal functioning of the plant is what we know as "disease", which can be caused by pathogenic organisms, such as fungi, bacteria, viruses, viroids, or nematodes, or by certain environmental conditions (Tronsmo et al., 2020).

What we observe as a plant disease is the result of the interaction between the host plant, the pathogen, and the environment (Fig. 3.1). This concept is known as the disease triangle and states that for disease to occur, a susceptible plant host, a virulent pathogen, and the proper environmental conditions are required, as a lack of favorable conditions for any of these three factors fails in the disease to develop (Velásquez et al., 2018).

3.1 Lifestyles of plant pathogens

Depending on how a pathogen acquires its nutrients from the host plant, it can be classified into three main lifestyles categories: (1) biotrophic, which usually are obligate parasites and can only obtain their nutrients from living host tissue; (2) necrotrophic, in which the pathogen kills the host cells to release

New Insights, Trends, and Challenges in the Development and Applications of Microbial Inoculants in Agriculture
https://doi.org/10.1016/B978-0-443-18855-8.00003-5

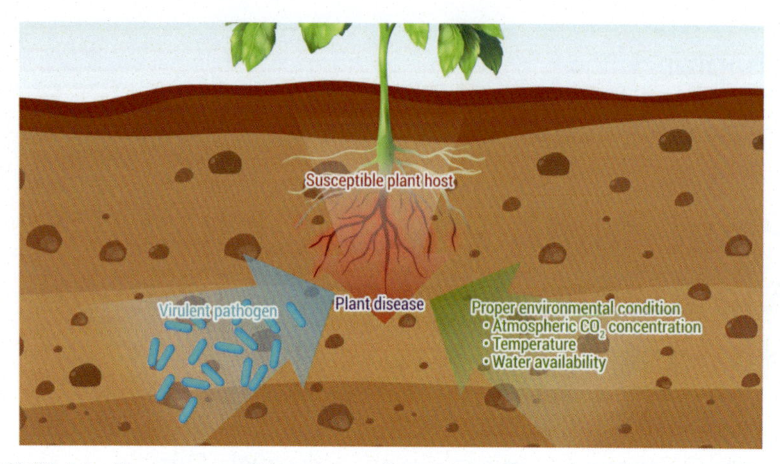

FIGURE 3.1 Necessary conditions for a plant disease. For a plant disease to occur, the presence of (i) a disease-susceptible plant, (ii) a virulent pathogen, and (iii) favorable environmental conditions are necessary.

nutrients; and (3) hemibiotrophic, which initially start as endophytes or biotrophs but progress to necrotrophy (Rodriguez-Moreno et al., 2017; Velásquez et al., 2018). Some examples of the mechanisms involved in the different lifestyles of plant pathogens are presented.

Phytopathogenic fungi that have a biotrophic phase use biotrophic invasive hyphae and secret effectors to alter the host defenses (Chowdhury et al., 2017; Mosquera et al., 2009). *Ustilago maydis* is a well-known fungal pathogen that causes smut in corn (*Zea mays*) and is used as a model for biotrophic interactions (Matei & Doehlemann, 2016). Its early biotrophic development happens intracellularly, where hyphae pass from cell to cell without causing visible symptoms. Then a biotrophic interaction zone is established to facilitate the exchange of nutrients and signaling molecules, including various protein factors; the fungus secretes effector proteins to suppress the plant's immune system and manipulate host metabolism (Basse et al., 2000; Djamei & Kahmann, 2012; Matei & Doehlemann, 2016). One of the effectors that facilitates breaking the plants' first defense barrier is the apoplastic effector Pep1, which suppresses pathogen-associate molecular pattern (PAMP)-induced immunity by inhibiting peroxidase POX12 activity (Hemetsberger et al., 2012). Another example is Tin2, which promotes disease progression by interacting with the maize protein kinase ZmTTK1, altering the lignin biosynthesis pathway toward an accumulation of anthocyanins, avoiding lignification of the tissue (Tanaka et al., 2014). For the acquisition of nutrients, biotrophic fungi have to divert the metabolism of the host to provide them with nutrients via the extracellular biotrophic interphase. *U. maydis* has been shown to express a plasma membrane-localized saccharose transporter (Srt1), which guarantees efficient carbon supply and transports the disaccharide saccharose without producing apoplastic signals that trigger plant defenses (Wahl et al., 2010). *Ustilago maydis* remains symptomless until the plant flowers and can infects all above-ground parts of the maize plant but fails to spread systemically and induces local tumors in which spores develop (Djamei & Kahmann, 2012).

There are a lot of examples of phytopathogens with a hemibiotrophic lifestyle, and one of the most economically important is the oomycete *Phytophthora* spp. In short, hemibiotrophic infection of *Phytophthora* consists of an early biotrophic phase, where specialized infection structures (haustoria) are formed to breach the plant cell walls and interface with the host membrane, after which rapid intercellular growth and colonization occur, ultimately leading to host cell death, sporulation, and initiation of a new infection cycle (Lamour et al., 2012). *Phytophthora capsici* is an important pathogen of tomato, other solanaceous, and cucurbit plants. Jupe et al. (2013) identified genes that mark specific infection stages of *P. capsici* and assessed their gene-expression profiles when infecting tomatoes. The expression of PcHmp1 (haustorial membrane protein 1), a marker for biotrophy, PcNpp1 (a necrosis-inducing protein known as Nep1-like protein 1), a marker for necrotrophy, and PcCdc14 (cell division control protein 14), a marker for sporulation, showed distinct expression patterns at different times during the infection of tomato, as determined by microarray and RT-PCR analysis. These patterns were consistent with stage-specific gene expression.

Necrotrophic phytopathogens kill host cells using toxic molecules and lytic enzymes, and they subsequently decompose the plant tissue and consume it for their growth. *Botrytis cinerea* (gray mold) is a model for necrotrophic pathogens with a broad host range, whose infection process comprises several stages (Prins et al., 2000). When *B. cinerea* lands on a leaf, it must penetrate the host surface and develop an appressorium, which requires a membrane-associated protein BcPLS1, which is homologous to a protein that is essential for appressorium function in *Magnaporthe grisea* (Gourgues et al., 2004). Upon breaching the cuticle, the penetration peg grows into the anticlinal wall of the underlying epidermal cell (Prins et al., 2000). The anticlinal cell wall is rich in pectin, and the early invasion of the epidermal cell layer must involve pectinases, specifically the endo-polygalacturonase BcPG2 (Kars et al., 2005). *Botrytis cinerea* possesses multiple tools that facilitate host cell death, such as phytotoxic metabolites (e.g., botrydial); the oxidative burst, triggered during cuticle penetration, and the secreted superoxide dismutase BcSOD1, which might play an important role; oxalic acid, a simple, versatile and potentially convenient molecule for pathogens; and host-selective toxins (van Kan, 2006). The infection by *B. cinerea* is promoted by and requires an active cell death program in the host, and several of the toxic compounds mentioned likely induce apoptosis, rather than causing disorganized necrosis (Van Baarlen et al., 2004; van Kan, 2006). Finally, *B. cinerea* decomposes and consumes the plant biomass; a common feature of all plant species that are colonized by *B. cinerea* is their high content of pectin. Therefore, it was postulated that this host preference reflects its possession of effective pectinolytic machinery (Kars et al., 2005; Wubben et al., 1999).

The interactions between pathogens and plants involve a complex network of pathogen-produced signals and the corresponding plant immune response regulators that intersect regardless of pathogen lifestyle. These networks and signaling mechanisms are described in the next sections.

3.2 Gene-for-gene in plant–pathogen interactions

Plants have evolved mechanisms that specifically recognize and respond to bacterial, fungal, and viral pathogen effectors. This interaction was observed between pathogens carrying single dominant genes (avirulence genes) that caused them to be recognized by plant hosts carrying single dominant resistance (R) genes, leading to the "gene-for-gene" name (Chisholm et al., 2006; Glazebrook, 2005). When pathogens are recognized and therefore fail to cause disease, they are called avirulent pathogens, the host is called resistant, and the interaction is called incompatible. This results in the initiation of defense signaling and host resistance. Resistance is manifested as localized cell death at the site of infection and inhibition of pathogen growth (Keen, 1990). Conversely, in the absence of gene-for-gene recognition, due to the absence of the avirulence gene in the pathogen and/or the absence of the R gene in the host, the pathogen is virulent, the host is susceptible, and the interaction is compatible, resulting in pathogen proliferation within the plant cells and the onset of disease (Chisholm et al., 2006; Keen, 1990). The R genes comprise several major groups, of which the largest are the nucleotide-binding leucine-rich repeat (NB-LRR) and the extracellular LRR (eLRR) resistance proteins (Chisholm et al., 2006; Glazebrook, 2005).

Based on the gene-for-gene mechanism, it is assumed that disease resistance conferred by R genes in tomatoes requires matching avirulence (AVR) genes in wilt fungus *Fusarium oxysporum* f. sp. *lycopersici* (Fol). The I (Immunity) gene encodes for a membrane-anchored leucine-rich repeat receptor-like protein (LRR-RLP) and is introgressed in tomatoes from *Solanum pimpinellifolium*, being one of the first plant disease resistance genes deployed against Fol race 1 (Bohn & Tucker, 1939), but its effectivity was short due to the emergence of Fol race 2 (Catanzariti et al., 2017). Similarly, the I-2 gene, also introgressed into a tomato from *S. pimpinellifolium* was deployed against Fol race 2 but, despite its greater durability, was eventually overcome by the emergence of Fol race 3 (Catanzariti et al., 2017). The I-3 gene was introgressed from *S. pennellii,* and later it was found that the matching AVR gene for the I-3 resistance gene encodes a secreted in xylem 1 protein (Six1), which is secreted by Fol during colonization of the xylem system and contributes to fungal virulence. Six1 is now called Avr3 to indicate its gene-for-gene relationship with the I-3 resistance gene (Rep et al., 2004). Races of Fol are named historically according to the R gene that is effective against them: the I gene and the I-1 gene are effective against race 1, race 2 overcomes I and I-1 but is stopped by I-2, while race 3 overcomes I, I-1, and I-2 but is blocked by I-3 (Houterman et al., 2008). This agricultural "arms race" between Fol and tomato is different from the naturally occurring one because it is dictated by successive R gene deployment in commercial cultivars.

3.3 Toxin production

Phytopathogenic fungus cultures provide a vast array of low molecular weight metabolites that have been demonstrated to be toxic to plants. Fungal toxins may disable host cellular functions or kill host cells. However, although it is easy to demonstrate that fungal cultures contain toxic substances, it is difficult to establish their role in plant disease. The development of DNA-based techniques for the transformation of fungal species during the 1980s provided a much-needed tool to test the role of toxins in plant pathogenesis (Desjardins & Hohn, 1997).

Fungal toxins can be grouped into two categories: (1) nonspecific toxins, which damage cells of phylogenetically unrelated plants, and (2) host-specific toxins, which are both determinants of pathogenicity and host range of a fungus (Idnurm & Howlett, 2001). Trichothecenes are nonspecific toxins produced by some *Fusarium* species that inhibit eukaryotic protein synthesis (Desjardins & Hohn, 1997). The role of trichothecenes in *Fusarium* pathogenesis has been investigated by generating trichothecene-nonproducing mutants through the disruption of Tri5, a gene controlling the first step of trichothecene biosynthesis. This resulted in reduced pathogenicity of *Giberella pulicaris* (anamorph *Fusarium sambucinum*) on parsnip but not in potatoes (Desjardins et al., 1996). The disruption of Tri5 also reduced the pathogenicity of *G. zeae* (anamorph *F. graminearum*) on some wheat cultivars in the laboratory and under field conditions (Desjardins et al., 1996; Proctor et al., 1995).

Host-specific toxins (HST) are molecules that are toxic only to the host of the disease and are innocuous to the great majority of other plants. Furthermore, only specific genotypes of the host, expressing a specific and often dominant susceptibility gene, are sensitive to the toxin (Friesen et al., 2008). *Alternaria alternata* pathotypes are characterized by this mechanism because of the morphological similarity between pathotypes but pathological differences (Tsuge et al., 2013). There are seven pathotypes of *A. alternata*, and each of their related HST has an essential role as a determinant of pathogenicity in all interactions between the plant host and *A. alternata*. These pathotypes and their HST are: (1) apple (*A. mali*), causing spot blotch through the AM-toxin, (2) Japanese pear (*A. kikuchiana*), which causes black spots and produces the AK-toxin, (3) rough lemon (*A. citri*), producing the ACR-toxin and causing leaf spot, (4) strawberry producing AF-toxin and causing black spot, (5) tangerine (*A. citri*), which causes brown spot with ACT-toxin, (6) tobacco (*A. longipes*), that also produces brown spot with AT-toxin, and (7) tomato (*A. alternata* f. sp. *lycopersici*), causing stem canker and producing the AAL-toxin (Akimitsu et al., 2014; Tsuge et al., 2013). The involvement of these toxins in pathogenicity has been demonstrated by the disruption of genes involved in the biosynthesis of AK-toxin (Tanaka et al., 1999) and of AM-toxin (Johnson et al., 2000), while the role of AAL-toxin as a pathogenicity factor on tomato was shown by random insertional mutagenesis (Akamatsu et al., 1997). In the case of AK-toxin biosynthesis, two linked genes (AKT1 and AKT2) were identified through insertional mutagenesis and targeted disruption of their flanking genes (AKTR-1 and AKT3-1) resulted in nonpathogenic strains (Tanaka et al., 1999; A. Tanaka & Tsuge, 2000).

3.4 Plant's defense triggered by pathogens: Hypersensitive response and systemic acquired resistance

The onset of a pathogen-caused disease triggers a series of defense mechanisms and pathways in the plant host, to stop, lessen, or counter the infection. The first line of defense for the host is its physical barriers, such as the waxy cuticle and active closure of stomatal pores; and depending on whether the pathogen is a virus, bacteria, or filamentous microorganism, it possesses mechanisms to break those barriers. For example, viruses can enter the plant by mechanical damage or by a biological vector, while filamentous fungi secrete cell-wall-degrading enzymes (CWDEs) (Silva et al., 2018). Pathogens that can overcome these barriers gain access to the host's apoplast, where the plant perceives and recognizes the pathogen as such, through transmembrane proteins called pattern recognition receptors (PRRs), that recognize specific conserved microbial features termed PAMPs (Castro-Moretti et al., 2020; Silva et al., 2018).

PAMP-triggered immunity (PTI) is the first active plant response to pathogen recognition. PTI consists of diverse cellular responses, including reactive oxygen and calcium bursts, mitogen-activated protein kinase (MAPK) signaling, plant hormone responses, transcriptional reprogramming, and cell wall fortification (Bigeard et al., 2015; Chisholm et al., 2006). One of the earliest physiological responses in PTI is an influx of extracellular Ca^{2+} into the cytosol (calcium burst), which induces the opening of other membrane transporters (influx of H^{+}, efflux of K^{+}, Cl^{-}, and NO_3^{-}), leading to an extracellular alkalinization and a depolarization of the plasma membrane (Bigeard et al., 2015). These changes are detected by Ca^{2+} sensors, such as calmodulin or calmodulin-like protein (CaM/CML), and calcium-dependent protein kinase (CDPK), and transduced into a signal, which leads to hypersensitive response (HR), a type of programmed cell death at the site of infection, as well as rapid production of H_2O_2 and NO (Ma et al., 2008). Another common plant response to different stresses and pathogens is the accelerated generation and accumulation of reactive oxygen species (ROS), including hydrogen peroxide (H_2O_2), superoxide anion, and hydroxyl radicals (Kovtun et al., 2000). H_2O_2 is an active signaling molecule, and its accumulation results in a variety of cellular responses, such as hypersensitive cell death (Lamb & Dixon, 1997), it blocks the cell cycle progression (Reichheld et al., 1999) and functions as a developmental signal for the onset of secondary wall differentiation; and it is also a potent activator of mitogen-activated protein kinases (MAPKs) in *Arabidopsis* leaf cells (Kovtun et al., 2000). MAPKs are conserved protein kinases in eukaryotes that establish signaling modules where MAPK kinase kinases (MAPKKKs) activate MAPK kinases (MAPKKs), which in turn activate MAPKs. In plants, upon detection of biotic or abiotic stress, MAPKs participate in the signal transduction to the nucleus, allowing adequate transcriptional reprogramming to occur during defense (Bigeard et al., 2015; Bigeard & Hirt, 2018). In plant immunity and defense, transcriptional reprogramming is a highly dynamic and controlled process, considered the main link between signal transduction (e.g., MAPK cascades) and implementation of induced defense mechanisms (e.g., production of antimicrobial compounds) (Bigeard et al., 2015; Moore et al., 2011).

The other two types of plant defense molecules, derived from secondary metabolites, are phytoalexins and phytoanticipins. The first ones are compounds that are produced by the plant host as a direct response to pathogen perception, whereas phytoanticipins are produced in advance of an attack and are only converted to their toxic forms post-pathogen perception, serving as a constitutive chemical barrier against the microbial attack (Castro-Moretti et al., 2020; Schlaeppi & Mauch, 2010).

Successful pathogens deploy a plethora of effectors that suppress PTI through susceptibility proteins (effector targets), allowing host cell infection and resulting in effector-triggered susceptibility (ETS). In response to effectors, plants developed a second line of receptors encoded by R genes that are activated via specific recognition of the effector or pathogen avirulence Avr proteins (see Section 3.2), resulting in effector-triggered immunity (ETI) (Silva et al., 2018). After recognition of a pathogen effector, plant R proteins induce robust plant defense that often includes the hypersensitive response (HR). The HR, associated with programmed death cells, withdraws essential nutrients from the pathogen and exposes it to a suite of antimicrobial compounds (Dangl,' et al., 1996). A local HR is often associated with the onset of systemic acquired resistance (SAR) in distal plant tissues, as signals derived from cells undergoing the HR contribute significantly to the induction of defense gene transcription in adjacent cells (Dangl,' et al., 1996; Khurana et al., 2005). These include mechanisms such as salicylic acid (SA) accumulation, pathogenesis-related (PR) gene expression, and broad-spectrum, long-lasting resistance to pathogen infection (Dempsey & Klessig, 2012).

The occurrence of HR cell death leads to the activation of SA signaling throughout the plant. After the accumulation of SA, a part is converted to methyl salicylate (MeSA), which acts as a phloem-mobile SAR signal (Dempsey & Klessig, 2012). MeSA can easily spread through systemic tissue; however, it alone does not induce defense gene expression. Once in systemic tissue, MeSA is modified to SA by the methyl esterase SABP2. This was observed in mutants that do not possess this enzyme and do not express PR genes in systemic leaves or develop SAR (Díaz-Puentes, 2012; Forouhar et al., 2005). SA levels rise, leading to the activation of various presumed defense effector genes, including PR genes (Glazebrook, 2005). SAR signaling downstream of SA is controlled by the redox-regulated protein Nonexpressor of PR gene 1 (NPR1), which, upon redox changes in the cell induced by SA, converts from its oligomeric form to monomers, then translocates from the cytosol into the nucleus, where it acts as a transcriptional coactivator for the TGA transcription factors to induce the expression of a large set of PR genes (Pieterse et al., 2014). PR proteins accumulate intercellularly and in vacuoles and have antimicrobial activity, as PR2 protein is a glucanase, PR3, and PR8 are chitinases, PR9 is a peroxidase, and PR1 has been shown to inhibit germination of oomycetes and have antifungal activity (Niderman et al., 1995). Disease resistance is given by diverse PR proteins that act together so that the overexpression or silencing of a single one does not have a major effect on the resistance or susceptibility of the plant to a range of pathogens (Díaz-Puentes, 2012).

All of these mechanisms, signaling, and metabolic pathways act together to stop pathogen infection in the host plant; thus, systemic acquired resistance is a mechanism of defense that confers long-lasting protection against a broad spectrum of microorganisms.

3.5 Disease-resistant plants

Plants have developed a wide array of defensive mechanisms underlying disease suppression to defend themselves from damage by pathogens. The response of plants to pathogen attack relies on pathogen recognition, which then triggers complex signaling pathways (discussed in Section 3.4). But at times, plants alone cannot resist the attack of virulent pathogens because their defense mechanisms are not strong enough to stop infection, environmental conditions are optimal for the disease development, and pathogens' effectors have evolved the ability to overcome host immune systems in the coevolution of host–microbe interactions (Chisholm et al., 2006).

Sustainable disease management is based on integral strategies, for example, the use of biological control agents (BCA) and plant growth-promoting microorganisms (PGPM), the use of plant extracts and metabolites, cultural practices, and the use of disease-resistant varieties, among others. The use of disease-resistant plants is an effective approach, which has extensively been employed to produce new disease-resistant varieties either through conventional breeding or genetic engineering (Nejat et al., 2017; Nelson et al., 2018; Pixley et al., 2019).

Plant pathogens have a versatile array of effectors that manipulate the host's immune system to benefit colonization. Characterization of core effectors, which are conserved among the population of a particular pathogen, is a promising strategy in disease resistance breeding, through identifying and editing of effector-target genes in the host, either executor genes or susceptibility genes (Nejat et al., 2017; Pixley et al., 2019). Turning on executor resistance genes by transcription activator-like effectors nucleases (TALENs) that bind sites upstream of resistance R genes can trigger disease resistance through a hypersensitive reaction (Strauβ et al., 2012). TALENs precisely induce DNA double-strand breaks (DSBs) at specific target positions in eukaryotic genomes. DSBs initiate natural DNA damage repair pathways either through nonhomologous end joining (NHEJ), which introduces small indels at the breakpoint, often resulting in frameshift mutation in the coding sequence that knockout a target gene, or homology-directed repair (HDR) as a template-dependent pathway accurately inserts the homologous donor template at the targeted genomic locus (Jankele & Svoboda, 2014). TALEN technology has successfully been used to precisely edit targeted genes in model plants and economically important crops such as rice (Christian et al., 2013; Li et al., 2012; Nejat et al., 2017; Zhang et al., 2013).

Clustered regularly interspaced short palindromic repeats (CRISPR) arrays are a diverse family of DNA repeats in ~40% of the sequenced bacterial and most archaeal genomes (~90%) (Horvath & Barrangou, 2010). The CRISPR-associated system (CRISPR/Cas) is an inheritable prokaryotic immune system that provides adaptive immunity against bacteriophages, conjugative plasmids, and foreign DNAs in a sequence-specific manner via RNA-guided target DNA silencing (Horvath & Barrangou, 2010). The Cas9 protein (a type II system of CRISPR/Cas) evolved as part of the bacterial immune system to destroy foreign nucleic acids, such as invading viral or plasmid DNAs (Nelson et al., 2018), and this function can be used in crop breeding to destroy invading pathogens in plants (Nejat et al., 2017; Nelson et al., 2018). *Citrus canker*, caused by *Xanthomonas citri* subsp. *citri*, has great damaging effects on the global citrus industry, and hence targeted editing of host disease-susceptibility genes represents an interesting alternative in citrus breeding for resistance (Nejat et al., 2017). A. Peng et al. (2017) reported an improvement in citrus canker resistance through CRISPR/Cas9-targeted modification of the susceptibility gene Lateral Organ Boundaries 1 (CsLOB1) promoter in Wanjincheng orange (*Citrus sinensis* Osbeck).

Although there is still much to be researched, strategies based on the regulation of the natural mechanisms of response to the attack of pathogens represent a viable and durable way to produce genetically modified plants, in contrast to strategies that introduce foreign and unique genes for resistance to pathogens, where the ability to confer resistance is quickly overcome by populations of pathogens (e.g., tomato I genes against Avr genes from *F. oxysporum* f. sp. *lycopersici*). Therefore, understanding the complex interactions that naturally occur between plants and pathogens is crucial to achieving the crop production needed to meet worldwide demand.

Chapter 4

Biological control agents for mitigating plant diseases

Amelia C. Montoya-Martínez[1], Valeria Valenzuela Ruiz[1], Roel Alejandro Chávez-Luzanía[1], Eber D. Villa-Rodríguez[2] and Sergio de los Santos Villalobos[1]
[1]*Instituto Tecnológico de Sonora, Ciudad Obregón, Sonora, Mexico;* [2]*Aarhus University, Aarhus, Denmark*

Food security and sustainability are global concerns; the growing population and consumer demand, the changing abiotic factors, and pathogens and pest resistance pressure have limited agricultural production (Barea, 2015). The economic and production losses caused by plant diseases vary depending to the plant species, as well as the pathogen, location, environment, and control measures implemented, or based on the combination of all these factors. These losses can range from minimum percentages to 100% (Agrios, 1995).

Integrated disease management aiming for more sustainable agriculture is a much-needed alternative to intensive conventional practices. This integrated management includes cultural practices such as crop rotations, reduced agrochemicals applications, and the use of plant growth-promoting bacteria (PGPB) as biofertilizers and biopesticides (Barea, 2015; Villarreal-Delgado et al., 2018). Among PGPB, those that protect, reduce, and control plant pathogens and pests are known as biological control agents (BCA). These are a diverse group of organisms that control plant diseases and pests through different mechanisms such as competition for space and nutrients, antibiosis, hyperparasitism, production of lytic enzymes and siderophores, and induction of plant resistance and priming (Fig. 4.1) (Köhl et al., 2019; Singh et al., 2020).

Bacterial BCA represent diverse genera, including *Agrobacterium*, *Arthrobacter*, *Azotobacter*, *Bacillus*, *Burkholderia*, *Collimonas*, *Paenibacillus*, *Pantoea*, *Pseudomonas*, *Serratia*, *Stenotrophomonas*, *Streptomyces*, and many others. Among these biocontrol bacteria, *Bacillus* and *Pseudomonas* are the most widely studied (Raaijmakers & Mazzola, 2012).

New Insights, Trends, and Challenges in the Development and Applications of Microbial Inoculants in Agriculture
https://doi.org/10.1016/B978-0-443-18855-8.00004-7

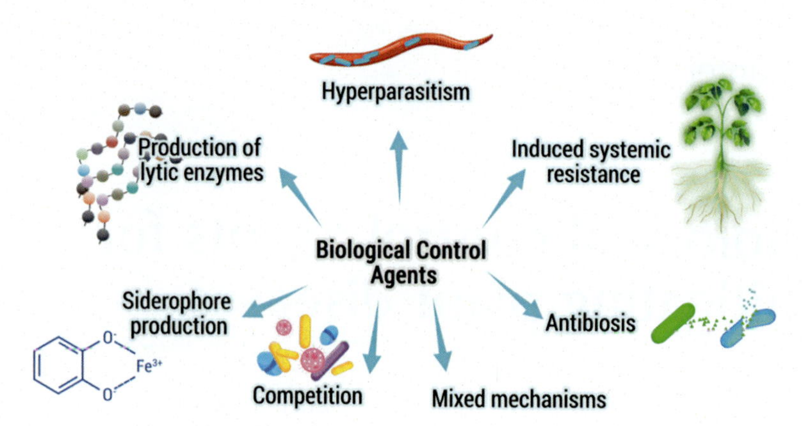

FIGURE 4.1 Mechanisms of biological control agents against phytopathogens.

4.1 Antibiosis

Although bacterial BCA have multiple mechanisms of biological control, the production of different kinds of antibiotics seems to be the principal mechanism of action against phytopathogens. These BCA can produce and secrete phloroglucinols, phenazines, pyoluteorin, pyrrolnitrin, rhamnolipids, and lipopeptides, among others (Köhl et al., 2019; Narayanasamy & Agents, 2013; Raaijmakers & Mazzola, 2012).

Fluorescent *Pseudomonas* spp. are abundantly found in the rhizosphere, and their ability to suppress the development of phytopathogens and associated diseases in a wide range of crops and ecosystems has been demonstrated. Strains of *Pseudomonas fluorescens* can produce 2,4-diacetyl phloroglucinol (2,4-DAPG) that inhibits the growth of various organisms, including fungi, bacteria, protists, and nematodes (Dehghaniana et al., 2019; Dutta et al., 2020). Several strains of fluorescent *Pseudomonas* spp. also produce antifungal metabolites such as phenazines, which comprise a large family of heterocyclic nitrogen-containing brightly colored pigments with broad-spectrum antibiotic activity (Masschelein et al., 2017; Meena, 2014; Narayanasamy & Agents, 2013). Phenazines derived from *P. chlororaphis* have been proven to be effective in the suppression of *Fusarium oxysporum* f. sp. *radicis-lycopersici* in tomato plants (Chin-a-woeng et al., 1998) and *Colletotrichum lindemuthianum* in bean (Bardas et al., 2009).

Similarly, *Bacillus* is a ubiquitous microorganism abundant in the *rhizosphere*, and the plant growth-promoting activity of some strains from this genus has been well-studied (Villarreal-Delgado et al., 2018). Strains of *Bacillus* can produce antimicrobial compounds such as surfactins, bacillomycin D, fengycins, azalomycin, and iturines (Athukorala & Fernando, 2009; Narayanasamy & Agents, 2013), with lipopeptides being the focus of many studies due to their antagonistic activity against a wide range of potential phytopathogens, including bacteria, fungi, and oomycetes (Valenzuela-Ruiz et al., 2020). In a comparative study involving multiple strains of *B. subtilis* and *B. amyloliquefaciens* tested against four different phytopathogens (*Cladosporium cucumerinum*, *Botrytis cinerea*, *Fusarium oxysporum,* and *Pythium aphanidermatum*), the data showed that the toxic activity of each lipopeptide family (*iturins*, *fengycins*, and *surfactins*) varies in function of the target species, and that the production of *iturins* and *fengycins* is modulated by the presence of pathogens (Cawoy et al., 2014). Recently, a study on the biological control of *Bipolaris sorokiniana* by *Bacillus cabrialesii* strain $TE3^T$ showed effective inhibition by the PGPB. Furthermore, genome-mining revealed that *B. cabrialesii* contains the biosynthetic potential to produce a wide spectrum of antimicrobial metabolites, and through bioactivity-guided LC-ESI-MS/MS approach, it was determined that a lipopeptide complex of *surfactin* and *fengycin* homologs was responsible for antifungal activity (Villa-Rodriguez et al., 2021a). Strains from the genera *Paenibacillus*, *Burkholderia*, *Serratia*, and *Streptomyces* can produce a wide array of antibiotics, including antifungal peptides (polymyxins), pyrrolnitrin, pyoluteorin, prodigiosin, and geldanamycin, among others, which have shown efficacy to control phytopathogens as *Rhizoctonia solani*, *Sclerotinia sclerotiorum*, *Penicillium expansum*, *Pythium* spp., *Colletotrichum graminicola*, *Fusarium* spp., and *Pseudomonas* spp. (Narayanasamy & Agents, 2013).

4.2 Production of lytic enzymes

Production of lytic enzymes by bacterial BCA might be exploited for biological control purposes, and it is a well-known mechanism for some strains of *Pseudomonas* spp., *Serratia marcescens*, *Paenibacillus* spp., *Bacillus* spp., *Streptomyces* spp., and *Lysobacter enzymogenes* (Kobayashi & Crouch, 2009; Singh et al., 2020). BCA parasitize fungal phytopathogens by excreting extracellular cell wall-degrading enzymes (CWDEs) such as chitinases, glucanases, and proteases that target the pathogen's cell wall, resulting in lysis of the phytopathogen cells. For example, *P. stutzeri* produces extracellular chitinase and laminarinase, which lyses the mycelia of *F. solani* (Lim et al., 1991). The enhancement of the natural CWDEs biosynthetic capacity of these BCA has been evaluated. For instance, the strain P5 of *P. fluorescens* that was transformed with the 6.5-kb chitinase gene fragment from *Serratia marcescens* strain M90-3, which showed inhibition of the mycelial growth of *Gaeumannomyces graminis* var. *tritici* and *Rhizoctonia solani*; the control by the transformant P5-1 was significantly greater than that of the wild-type strain (Xiao-Jing et al., 2005). *Bacillus cereus* has been evaluated for its ability to produce lytic enzymes and it was found that it can secrete a complex of hydrolytic enzymes such as chitinase, chitosanase, and protease when grown in a medium containing chitosan flakes from marine waste. Furthermore, the culture supernatant significantly inhibited the growth of *F. oxysporum*, *F. solani*, and *Pythium ultimum* (Chang et al., 2009).

As its name implies, *Lysobacter enzymogenes* produces a variety of lytic enzymes, such as chitinases, β-1,3-glucanases, and proteases, all of which are capable of degrading cell wall components of fungi and stramenopiles. The role of lytic enzymes in the biological control activity of *L. enzymogenes* strain C3 has been studied by evaluating the purified chitinolytic fractions produced by the strain, which inhibited spore germination and infection in turfgrass caused by *Bipolaris sorokiniana* (Zhang et al., 2001). *Lysobacter enzymogenes* has also been shown to have biological control against *Pythium* sp., *Aphanomyces cochlioides*, *Fusarium graminearum*, *Rhizoctonia solani*, *Sclerotinia sclerotiorum*, and others (Kobayashi & Crouch, 2009; Li et al., 2008; Narayanasamy & Agents, 2013). The biological control potential of *Serratia marcescens*, *Streptomyces viridodiasticus*, and *Micromonospora carbonacea* against *Sclerotinia minor* was evaluated, and it was found that these bacteria and actinomycete effectively suppressed phytopathogen growth and disease incidence. These BCA were efficient producers of chitinase and β-1,3-glucanase and caused extensive plasmolysis and cell wall lysis of *S. minor* (El-Tarabily et al., 2000).

4.3 Competition for space and nutrients

Competition for niche and/or nutrients is a control mechanism in which the BCA can occupy a niche and rapidly consume essential nutrients for phytopathogen infection, so that the outcompeted pathogens are unable to infect the host plant and, consequently, the disease is suppressed (Köhl et al., 2019). The competition for a niche depends on the effective and efficient colonization of the BCA in the rhizosphere and host tissue. Colonization by BCA depends on several factors, of which motility plays an important trait. In *Pseudomonas fluorescens* F113, motility is a polygenic trait, and phenotypic variants of strain F113, which showed translucent and diffuse colony morphology, had enhanced colonization of alfalfa rhizosphere. In addition, all the isolated variants were more motile and competitive than the wild-type strain, displacing it from the root tip within 2 weeks. Biocontrol experiments with phenotypic variants of F113 against *Fusarium oxysporum* f. sp. *radicis-lycopersici* and *Phytophthora cactorum* have shown that these strains possess biocontrol activity (Barahona et al., 2011; Martínez-Granero et al., 2006).

Bacillus subtilis EXWB1 has shown to have a unique mechanism of biocontrol activity against *Alternaria alternata*, fungi that infect melon fruits during transport and storage. The droplet of the strain EXWB1, when placed on melon skin, spread as a thin film on the hydrophobic fruit surface, indicating that the bacterial strain established effective adhesion to the fruit surface, suggesting that the bacterial strain could rapidly occupy the same niche as the fungal pathogen and the hyphae of *A. alternata* could not grow on melon skin inoculated with EXWB1 and no rotting of the fruit flesh could be observed in fruits treated with bacterial cell suspension (Wang et al., 2010).

4.4 Production of siderophores

The availability of essential micronutrients, such as iron, is a crucial factor for the growth of all microorganisms. Therefore, microorganisms that have the ability to produce high amounts of siderophores with a high affinity to iron play an important role in disease suppression and can be selected for biological control by competing with phytopathogens that produce lesser amounts of siderophores with lower affinity for iron (Köhl et al., 2019; Lugtenberg & Kamilova, 2009; van Loon, 2000). The fluorescence of *Pseudomonas* spp. is attributed to the presence of an extracellular pigment called pyoverdine (Pvd) or pseudobactin. This pigment has a high affinity for iron ions and acts as a siderophore (Narayanasamy & Agents, 2013). Bacterial isolates obtained from soil samples were tested for their siderophore production and effectiveness in inhibiting the mycelial growth of *Alternaria* sp., *Fusarium oxysporum*, *Magnaporthe grisea*, *Pyricularia oryzae*, and *Sclerotium* sp., which infect rice. It was found that 23% of the isolated bacteria produced siderophores and, in dual-culture technique, siderophore-producing bacteria showed strong antagonistic activity against all the five rice pathogens to varying degrees, ranging from 10.4% to 37.5%. *Streptomyces* sp. and *Pseudomonas* sp. showed significantly higher antagonistic effect against *Alternaria* sp., while *Bacillus firmus* inhibited *P. oryzae*. In addition, *P. aureofaciens* was the best siderophore producer overall, exhibiting an in vitro antagonistic effect against *Alternaria* sp., *F. oxysporum*, and *P. oryzae* (Chaiharn et al., 2009).

4.5 Hyperparasitism

Parasitism is the direct competitive interaction between two organisms in which one organism gains nutrients from the other. If the host is also a parasite, such as a plant pathogen, the interaction is defined as hyperparasitism (Köhl et al., 2019). Hyperparasites usually kill and invade spores, mycelium, cells of bacterial pathogens, and endospores of fungal pathogens (Singh et al., 2020). This kind of interaction is often observed between fungi. However, hyperparasitism is rarely reported for bacteria, and one particular example of bacteria hyperparasitizing other bacteria is the small deltaproteobacteria *Bdellovibrio bacteriovorus*. It has evolved a novel lifestyle whereby it invades the periplasm of other gram-negative bacteria and digests and kills them from the inside, using the cytoplasm of other bacteria as nutrients (McNeely et al., 2016; Negus et al., 2017). This predator is ubiquitous and can be found in a variety of habitats such as soil, sewage, freshwater, and marine environments (Rotem et al., 2014). The characteristic predatory mechanism of *Bdellovibrio bacteriovorus* has been explored for biocontrol purposes, but the effectiveness of these predators as biocontrol agents relies on their prey range. In this sense, the prey range of three isolates (DM7C, DM8A, and DM11A, identified as *Bd. bacteriovorus*) was tested on *Burkholderia cepacia* complex bacteria and several phytopathogenic bacteria of agricultural importance (*Agrobacterium tumefaciens*, *Xanthomonas vesicatoria*, *X. campestris* pv. *campestris*, *Erwinia carotovora* pv. *carotovora*, *Erwinia herbicola*, *Pseudomonas syringae* pv. *glycinea*, *P. syringae* pv. *Tomato* and *P. marginalis*). It was found that the *Bd. bacteriovorus* strains did not distinguish their prey range between isolates of *B. cepacia* complex or the other tested phytopathogens. Such strains hold promise as potential wide-spectrum biocontrol agents (McNeely et al., 2016).

Another example of hyperparasitic bacteria is the phytoparasitic nematode predator, *Pasteuria* spp. The *Pasteuria* genus is of particular interest as these are obligate, mycelial, endospore-forming bacterial hyperparasites of phytoparasitic nematodes (Tian et al., 2007). The spores of *Pasteuria* spp. attach to the cuticles of juvenile root-knot nematodes (as *Meloidogyne* spp.), to germinate on and penetrate the cuticle, where vegetative microcolonies then form and proliferate inside the body of the developing female; due to reproductive system degeneration, the adult female is almost devoid of eggs (Gowen et al., 2007; Tian et al., 2007). Bhuiyan et al. (2018) established an experiment where sugarcane was grown in pasteurized sand containing different concentrations of endospores of *P. penetrans* for the control of *M. javanica* showing that the severity of root galling and the number of nematode eggs decreased as the endospore concentration increased. This genus has proven its potential as a BCA against phytoparasitic nematodes, however, the obligate nature of the bacteria life cycle has made it difficult to mass produce it in vitro (Gowen et al., 2007).

4.6 Inducing resistance and priming of the plant

Interaction of some BCA with plant roots can result in plants becoming resistant to some pathogenic bacteria, fungi, and viruses. This mechanism is known as induced systemic resistance (ISR). ISR has emerged as an important mechanism by which PGPB prime the whole plant for enhanced defense against a broad range of pathogens. Screening for BCA with high potential in inducing resistance is complex because the mode of action depends on a sequence of events, including the establishment of the BCA, the release of signaling compounds, the induction of a cascade of metabolic events to induce plant defense mechanisms, and the response of the pathogen to these defense mechanisms (Köhl et al., 2019; Pieterse et al., 2014; Van Loon & Bakker, 2005). ISR can be triggered through diverse compounds produced by the bacterial BCA, known as elicitors. Cell surface components, such as the outer membrane lipopolysaccharide and flagella can elicit ISR. Siderophores produced by *Pseudomonas aeruginosa*, *P. fluorescens*, and *Serratia marcescens* can also act as ISR elicitors. Antibiotics and volatile compounds such as 2,3-butanediol have also been implicated in ISR (Van Loon & Bakker, 2005). In one of the first ISR studies on the antagonistic activity of *Pseudomonas fluorescens* strain WCS417 against *Fusarium oxysporum* f. sp. *dianthi* on carnation, it was found that the bacteria, when confined to the plant root system, remained protective even when the pathogen was slash-inoculated into the stem. Since, in this case, the BCA and the phytopathogenic fungus never came into contact on the plant, the protective effect had to be plant-mediated (Peer, 1991). This topic is discussed in more detail in Chapter 7.

4.7 Mixed mechanisms of biological control

Biological control is an important tool for sustainable agriculture, and the understanding of the mechanisms of BCA involved in it is crucial for exploiting their full potential. As mentioned before, the main biocontrol mechanisms include antibiosis, production of lytic enzymes, competition, hyperparasitism, and induced resistance. While individual biocontrol mechanisms are predominant for some BCA, there are also many instances where multiple mechanisms may operate in a given BCA isolate. For example, the antibiotic DAPG derived from *Pseudomonas fluorescens*, previously mentioned, can have a direct effect as an antimicrobial metabolite against pathogens and also acts as an elicitor for ISR in the plant (De Vleesschauwer & Höfte, 2009; Pieterse et al., 2014). Thus, both antibiosis and induced resistance act simultaneously, and artificial separation between the effect of DAPG on a single mechanism is hardly possible. One of the most conclusive pieces of evidence for the involvement of antibiotics in BCA-triggered systemic resistance was provided by Iavicoli et al. (2003), where it was demonstrated that DAPG has a determinant role in the induction of systemic resistance in *Arabidopsis thaliana*. In their findings, DAPG produced by *P. fluorescens* CHA0 was shown to induce resistance against the oomycete *Hyaloperonospora parasitica*, whereas only the mutations interfering with DAPG production led to a significant decrease in ISR (Iavicoli et al., 2003).

Another example is the production of siderophores for nutrient competition with the pathogen, which are also recognized by plants, thus inducing systemic resistance (De Vleesschauwer et al., 2008; De Vleesschauwer & Höfte, 2009). In 2008, De Vleesschauwer et al. demonstrated the ability of *Pseudomonas fluorescens* WCS374r to trigger ISR in rice against the pathogen *Magnaporthe oryzae*, showing that this WCS374r-induced resistance is regulated by an independent salicylic acid (SA) but jasmonic acid/ethylene-modulated signal transduction pathway. Moreover, through bacterial mutant analysis a pseudobactin-type siderophore was identified as the crucial determinant responsible for ISR elicitation.

Analyzing and considering a single mechanism of action of BCA is a scientific exercise to unravel how BCA act, and this information is important for optimizing their use. However, the nature of microbial interactions is more complex and does not fit into such a pragmatic point of view. In many cases where the mechanism has been extensively studied for a particular biocontrol strain, the results confirm that antagonistic interactions are driven by more than one mode of action (Köhl et al., 2019).

4.8 Combined use of multiple BCA

In recent years, there has been increasing interest among researchers in using mixtures of BCA. A variety of biocontrol mechanisms may operate in mixed BCA populations, but direct and indirect interactions between the different BCA need to be taken into account. Compared with the most effective BCA, the combined use of two BCA may lead to higher, lower, or similar biocontrol efficacy. Recent theoretical modeling work suggests that disease suppression by the combined use of two BCA is in general, very similar to that achieved by the more effective one, indicating no synergistic interactions (Xu et al., 2011).

A necessary precondition for the success of a BCA mixture or consortium is the analysis of the microbial interactions among its members and the effect exerted on plant health, considering compatibility and diversity (Niu et al., 2020). Compatibility of BCA in a consortium refers to their ability to coexist without inhibiting the growth of each other during their in vitro co-culture and/or in rhizosphere colonization competition assays (Thomloudi et al., 2019). In addition, the degree of microbial diversity affects the assembly, survival, and functionality of BCA in the rhizosphere and their ability to inhibit diseases (Hu et al., 2016).

For example, a bacterial consortium consisting of seven strains, *Enterobacter ludwigii*, *Stenotrophomonas maltophilia*, *Ochrobactrum pituitosum*, *Herbaspirillum frisingense*, *Pseudomonas putida*, *Curtobacterium pusillum*, and *Chryseobacterium indologenes*, representing three of the four most dominant phyla found in maize roots, was able to reduce the severity of maize seedling blight caused by *Fusarium verticillioides*, and it was found that the biocontrol effect of the mixed BCA consortium was stronger than that of each individual strain (Niu et al., 2017).

In the current scenario of crop production, biological control is of utmost importance; however, its application needs to be optimized. The use of emerging strategies for improving BCA benefits can help to achieve this goal. These strategies include the use of plant exudates to attract beneficial bacterial BCA, as certain PGPB are attracted by specific root exudates to meet specific needs. Also, the use of substrates to maintain beneficial BCA populations and the breeding of plants optimized to attract and maintain the colonization of beneficial microbes are prime objectives of this approach (Hussain et al., 2020). Progress in molecular approaches has escalated the potential for improving the properties of BCA. Biotechnological applications, such as genetic engineering, allow modification of BCA properties to enhance their survival under stress and nutrient-limited conditions, enhance the production of antibiotic compounds, grow under varied pH and temperature conditions compared to the original strain (Singh et al., 2020).

Bacterial biological control agents use a broad array of mechanisms that occur when organisms interact, communicate, and regulate their coexistence in their plant-pathogen-BCA interactions. The exploitation of different modes of action has different advantages and disadvantages for the development of commercial BCA inoculants. Microbial inoculants based on bacterial BCA, such as *Bacillus* and *Pseudomonas*, are promising because they not only suppress phytopathogens but also are known for their plant growth promotion benefits. Bacterial BCA are fundamental for achieving effective integrated disease and pest management.

Chapter 5

Plant growth-promoting bacteria as a sustainable agricultural strategy

Alondra María Díaz-Rodríguez, Roel Alejandro Chávez-Luzanía, Amelia C. Montoya-Martínez and Sergio de los Santos Villalobos
Instituto Tecnológico de Sonora, Ciudad Obregón, Sonora, Mexico

Plant growth-promoting bacteria (PGPB) are beneficial bacteria that colonize plant tissues and roots, stimulate growth, improve nutrient availability in the soil, protect plants from diseases, and enhance plant tolerance to abiotic stress. PGPB include those that are free-living, symbiotic bacteria, rhizospheric bacteria, and endophytes (Ambrosini et al., 2016).

The density and diversity of the PGPB in the plant and rhizosphere can vary depending on the plant species, crop age, season, soil physicochemical characteristics, and agricultural practices (Shameer & Prasad, 2018). Nonetheless, the application of PGPB has been proven to be an environmental strategy for improving plant growth and development through direct and/or indirect mechanisms. Direct mechanisms refer to bacterial traits that directly promote plant growth by sequestration, solubilization, or fixation of nutrients, as well as by the production of growth regulators that influence hormone levels in plants. On the other hand, indirect mechanisms assist the plant to develop healthier in adverse environmental conditions (Glick, 2012; Gupta et al., 2015).

5.1 Direct mechanisms

The microbial direct mechanisms include the (1) modulation of plant hormones, (2) enzyme production, such as 1-aminocyclopropane-1-carboxylate (ACC) deaminase, and (3) facilitation of nutrient acquisition, including biological nitrogen fixation, phosphorus solubilization, and iron sequestration (Fig. 5.1) (Ferreira et al., 2019; Gupta et al., 2015; Santoyo et al., 2019).

New Insights, Trends, and Challenges in the Development and Applications of Microbial Inoculants in Agriculture
https://doi.org/10.1016/B978-0-443-18855-8.00005-9

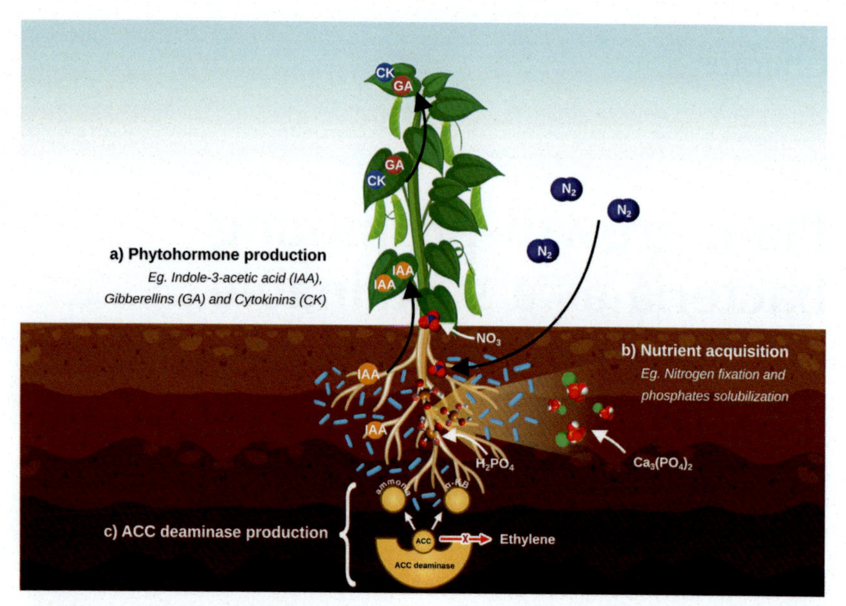

FIGURE 5.1 **Direct mechanisms of PGPB.** Plant growth-promoting bacteria can carry out beneficial effects on plants through direct and indirect mechanisms. Some examples of direct mechanisms of microorganisms are (A) the modulation of plant hormones such as auxin indole acetic acid (IAA), gibberellins (GA), and cytokinins (CK), which regulate various processes during the growth and development of the plant; (B) the acquisition of essential nutrients such nitrogen and phosphorus, which are often limited in the soil for crop and forage production; and (C) the ability of microorganisms to alleviate different types of stress by the enzyme 1-aminocyclopropane-1-carboxylate (ACC) deaminase.

5.1.1 Phytohormones modulation/production

Plant hormones are compounds produced by plants to regulate and control plant growth that provide their main effect at a cellular level at low concentrations. These regulators can also be synthesized by other organisms, often with greater potency than their plant counterparts. This characteristic can have either beneficial or inhibitory effects on the crop in question (Romera et al., 2019). One of the beneficial effects that plants can obtain is through the regulation of phytohormone levels by the action of PGPB. PGPB have the ability to synthesize and secrete compounds such as auxins, gibberellins (GA), cytokinins (CK), abscisic acid (ABA), acid salicylic acid (SA) jasmonic (JA), and ethylene (ET) (Etesami & Maheshwari, 2018; Romera et al., 2019). These phytohormones have been extensively studied for their potential to improve plant growth conditions, specifically by triggering various adaptive responses including the mediation of growth and development, nutrient allocation, and source/sink transitions (Bari & Jones, 2009; Messing et al., 2010).

Auxins are one of the most common phytohormones in the rhizosphere, particularly indole-3-acetic acid (IAA) which is produced by 80% of rhizospheric bacteria. The production of IAA acts as a reciprocal signaling molecule between PGPB and plants. Its presence can lead to several beneficial effects, including (1) increased roots area, which contributes to more efficient water and nutrient uptake; (2) stimulation of tuber and seed germination; (3) regulation of various plant growth processes; (4) enhancement of xylem and root development rates; (5) initiation of lateral and adventitious root formation, and (6) improvement of plant pigment production and photosynthetic rate (Etesami et al., 2015; Gupta et al., 2015; Spaepen & Vanderleyden, 2011).

Gibberellins are phytohormones belonging to the diterpenoid family, including C_{20}-GAs and C_{19}-GAs (Salazar-Cerezo et al., 2018). These biomolecules have been identified as regulators of various processes of plant development (Briones-Moreno et al., 2017). They can positively regulate the gene expression for gibberellin synthesis, increasing the production of gibberellin receptors, and enzymes that catabolize bioactive gibberellins resulting in plant growth promotion and overcoming biochemical repressions that plants may experience (Salazar-Cerezo et al., 2018). On the other hand, gibberellins have been used to increase the number of flower buds in different plant species, such as *Rhododendron pulchrum* (Chang & Sung, 2000), *Tulipa gesneriana* L. (Ramesh et al., 2013), black iris (Al-Khassawneh et al., 2006), carnation (Ibrahim, 2017), and sweet pea (Asil et al., 2011). In addition, they are also used to reduce undesired defects in fruits such as apples and pears (Knoche et al., 2011).

Cytokinins are involved in various plant growth processes, including cell division, photosynthesis, chloroplast differentiation, and regulation of leaf senescence (Li et al., 2021). They also have the ability to increase root biomass, allowing plants to develop under drought conditions (Arkhipova et al., 2007; Liu et al., 2013). Studies on the overexpression of A-type Arabidopsis Response Regulators (ARRs) genes involved in plant development, have shown that the addition of cytokinins regulates the transcription of ARRs, resulting in growth promotion, specifically of the roots (Ren et al., 2009). It has been observed that the levels of cytokinins produced by PGPB (regularly low) produce beneficial effects on plants, however, even though some phytopathogens produce cytokinins, these secrete them in high concentrations, which act as inhibitors of plant growth (Gupta et al., 2015).

The production of ethylene by the plant itself and the PGPB present in the rhizosphere serves as an activator of growth-promoting processes, such as root elongation, fruit ripening, reduced leaf wilting, increased seed germination, promotion of leaf abscission, and the production of other plant hormones (Gupta et al., 2015). As mentioned, the beneficial effects of phytohormones are observed at low concentrations, which is why high ET concentrations induce defoliation and other metabolic processes that reduce plant yield (Bhattacharyya & Jha, 2012).

5.1.2 1-Aminocyclopropane-1-carboxylate deaminase

The natural presence of ethylene in plants plays a crucial role in their development, and it is present in almost all tissues and stages of plant growth, in addition, it participates in combating different types of stress (Olanrewaju et al., 2017). However, ethylene synthesis is also influenced by stress conditions caused by abiotic factors such as drought, flooding, temperature, or contamination by heavy metals, as well as biotic factors such as attacks by phytopathogens. While ethylene can help plants combat these stress conditions, high concentrations of this hormone can lead to chlorosis, abscission, and senescence (Santoyo et al., 2019).

Beneficial microorganisms that come into contact with plants can alleviate this stress condition by synthesizing the enzyme ACC deaminase. This enzyme can hydrolyze the ethylene precursor, 1-aminocyclopropane-1-carboxylate (ACC), converting it to α-ketobutyrate and ammonia. By interrupting the ACC metabolic pathway, it effectively decreases ethylene concentrations in plants (Parra-Cota et al., 2018; Santoyo et al., 2019). Several species, particularly the genera *Pseudomonas*, *Bacillus*, *Achromobacter*, *Alcaligenes*, *Burkholderia*, *Enterobacter*, *Kluyvera*, *Sinorhizobium*, and *Variovorax*, have been reported to possess high capacities to mitigate the negative impacts of stress in plants through the synthesis of ACC deaminase (Compant et al., 2019; Etesami & Maheshwari, 2018; Hussain et al., 2020).

5.1.3 Nitrogen fixation

Nitrogen (N) is one of the precursors of proteins, nucleic acids, and essential biomolecules in all known life forms. However, nitrogen is not a naturally abundant element in the soil, at least not in forms useable by plants. Processes such as soil erosion, desertification, leaching, and chemical volatilization contribute to the constant loss of this essential element (Soumare et al., 2020). Due to its great importance and scarcity of bioavailable nitrogen in the soil, this nutritional requirement of plants is usually supplied through the constant incorporation of nitrogen fertilizers into agricultural soils. However, this practice implies a high economic and environmental cost (Yadav, 2020).

Although Earth's atmosphere is composed of 78% of N (making it the main source of this element on Earth), it is found in the form of gaseous dinitrogen, which cannot be directly used by plants (Ferguson et al., 2010). Biological nitrogen fixation by PGPB is carried out by a dehydrogenase enzyme that converts atmospheric dinitrogen into ammonium. This process consists of a combination of two dinitrogen molecules with the hydrogen ions of a water molecule, a mechanism that is highly conserved among plant growth-promoting microorganisms (PGPM) (Soumare et al., 2020). Notably, this mechanism is exclusive to prokaryotes, among which the following genera stand out *Azotobacter*, *Azospirillum*, *Bacillus*, and *Clostridium*, qualified as free-living, and symbiotic bacteria such as *Rhizobium*, *Frankia*, and *Cyanobacteria* (Ininbergs et al., 2011; Ravikumar et al., 2007). This process is the primary means in which plants obtain nitrogen, and plays a crucial role in the distribution of this element throughout the ecosystem (Soumare et al., 2020).

Biological nitrogen fixation is particularly efficient in the rhizobial group, which comprises microorganisms such as *Azorhizobium*, *Allorhizobium*, *Bradyrhizobium*, *Mesorhizobium*, *Rhizobium*, and *Sinorhizobium*. These PGPB engage in a symbiotic relationship with plants, in which they fix atmospheric nitrogen in exchange for carbon sources provided by the plant to support microbial energy production and subsequent proliferation (Ferguson et al., 2010). As mentioned above, nitrogen is a limiting element in crop productivity, so the ability of legumes to establish strong symbiotic associations with nitrogen-fixing PGPB provides a viable alternative to nitrogen fertilizers.

5.1.4 Phosphate solubilization

After nitrogen, phosphorus (P) is the most important element in plant nutrition, constituting approximately 0.2% of the total dry weight of plants (Beltrán-Pineda, 2014) and is involved in essential processes, such as photosynthesis, energy transfer, cellular respiration, and biomolecule synthesis (Gupta et al., 2015). However, the fundamental role played by phosphorus as a plant nutrient is hindered by its high retention in the soil, low mobility, and limited replenishment for continuous root absorption. Therefore, assistance in obtaining this element by PGPB is one of the most important indirect mechanisms for promoting plant growth (Gupta et al., 2015; Valenzuela-Ruiz et al., 2021).

Plants can only take up phosphorus in two of its ionic forms: $H_2PO_4^{1-}$ and HPO_4^{2-}. Thus, it is crucial to find effective ways for crops to acquire this element (Beltrán-Pineda, 2014; Bhattacharyya & Jha, 2012). To meet the nutritional needs of crops, excessive amounts of phosphorus fertilizers are often applied due to the limited presence of bioavailable phosphorus. However, it is known that excessive use of these agrochemicals contributes to the eutrophication of water bodies and around 90% of the precipitated phosphorus forms compounds that are not usable by plants.

The root systems of plants host PGPB that perform different mechanisms to solubilize phosphate compounds that cannot be used by plants, leaving P in a mobile and soluble state so that crops can obtain this element, which in turn provides carbon compounds used for microbial growth (Beltrán-Pineda, 2014). The mechanisms of phosphorus solubilization involve the production of extracellular mineral solvents (organic acid anions, protons, hydroxyl ions, and CO_2), as well as the release of enzymes (biochemical mineralization), and substrate degradation leading to the release of P (biological mineralization) (Díaz-Rodríguez et al., 2021; Sharma et al., 2013). However, the main mechanism of phosphate solubilization is the decrease of soil pH to levels close to 2, this occurs through the expression of organic acids by microorganisms in the rhizosphere. This acidification promotes the formation of insoluble complexes with other metals, thus releasing the phosphorus present in soil (Fernández et al., 2005).

Research has revealed that interactions between mycorrhizal fungi with certain plant growth-promoting rhizobacteria enhance their activity in the rhizosphere, leading to increased absorption of phosphorus and other nutrients such as zinc (Zn) and copper (Cu). This interaction, in the specific case of phosphorus, allows the mycelium of mycorrhizae to take up phosphate released by the rhizobacteria, resulting in an overall improvement in the acquisition of this element by the plant (Wahid et al., 2020).

5.2 Indirect mechanisms

On the other hand, indirect mechanisms, which are related to the protection of plants from biotic (infections) or abiotic (environmental) stresses, include (1) production of antibiotic compounds, (2) lytic enzyme production such as lipopeptides, chitinases, β-glucanases, and cellulases, (3) siderophore production, (4) competition for nutrients and space, and (5) endotoxins production (Shameer & Prasad, 2018; Villarreal-Delgado et al., 2018). Furthermore, growth promotion could happen by modulating the effects of environmental stress through (1) IAA, (2) ethylene, (3) cytokinin, (4) trehalose, (5) antioxidant enzyme activities, and (5) antifreeze proteins (Fig. 5.2) (Egamberdieva et al., 2017; Glick, 2012; Shameer & Prasad, 2018).

5.2.1 Antibiotic production

One of the most studied and powerful mechanisms of pathogen growth inhibition is the production of antibiotics by PGPB and biological control agents (BCA), which is an important direct inhibition strategy for combating plant diseases (Qu et al., 2020). These antibiotics are oligopeptides that inhibit the synthesis of pathogen cell walls, influence membrane structures of cells, and inhibit the formation of the initiation complex on the small subunit of ribosomes (Maksimov et al., 2011). To date, several studies have reported six different groups of antibiotics, including phloroglucinols, phenazines, pyoluteorin, cyclic lipopeptides, and pyrrolnitrin, which all of these are hydrogen cyanide, volatile, and diffusible, mostly associated with the biological control of various plant infections (Hakim et al., 2021; Mitra et al., 2020).

The antibiotic activity is primarily attributed to the vascularization of the protoplasm, pore formation, and complete disintegration of the cells. In most cases, the cell membrane is the target of their activity, specifically, cationic peptides induce the formation of channels through which ions can pass and disrupt the cell membrane (Sumi et al., 2015). Among antibiotic-producing microorganisms, *Streptomyces* sp. contributes to the production of almost

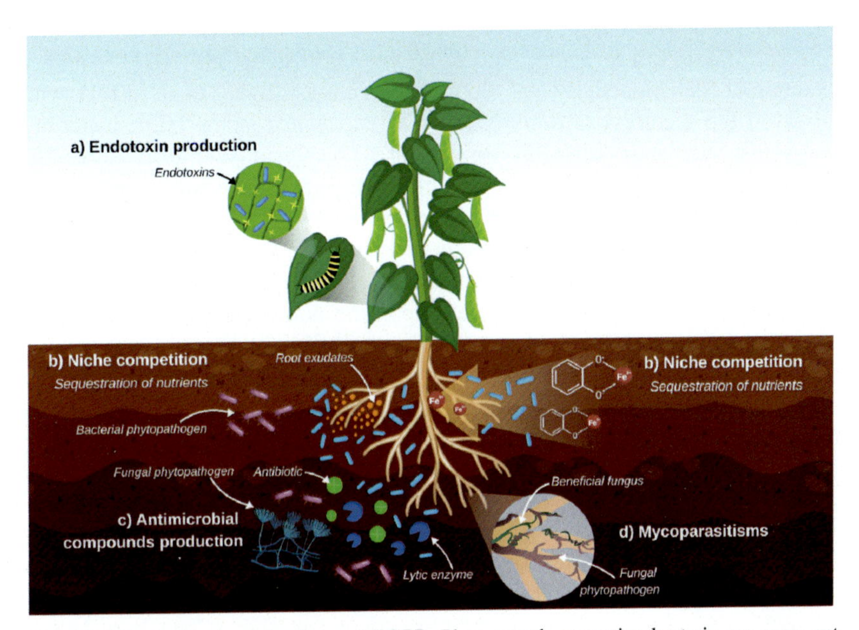

FIGURE 5.2 Indirect mechanisms of PGPB. Plant growth promoting bacteria can carry out beneficial effects on plants through direct and indirect mechanisms. Regarding the indirect mechanisms, some microorganisms can (A) produce endotoxins, (B) compete with the pathogens to obtain the nutrients that exude from the roots of the plants and occupy the best spaces and niches, (C) synthesize an array of antimicrobial compounds and enzymes that exhibit lytic activity to inhibit, restrict, or eliminate the growth of various phytopathogens, and (D) direct attack of fungal pathogens by mycoparasitism.

80% of the total antibiotics known to date (Ahmed et al., 2021). However, other well-studied and reported genera capable of antibiotic production include many *Bacillus* species (Thakur et al., 2021), *Pseudomonas* spp., *Stenotrophomonas* spp., *Agrobacterium* spp., and *Burkholderia. Besides,* many other pathogenic fungi such as *Alternaria solani*, *Aspergillus flavus*, *Botryosphaeria ribis*, *Colletotrichum gloeosporioides*, *Fusarium oxysporum*, *Helminthosporium maydis*, and *Phomopsis gossypi*i have also been identified as antibiotic producers (Mitra et al., 2020).

For example, among the 167 antibiotic types produced by *Bacillus*, more than 12 are synthesized by *B. subtilis* strains, including bacillomicin, mycobacillin, fungistatin, iturin, fengycin, plipastatin, surfactin, and bacilizin, among others (Maksimov et al., 2011). The majority of Bacillus antibiotics (e.g., polymyxin, circulin, and colistin) exhibit activity against both gram-positive and gram-negative bacteria, and pathogenic fungi such as Alt*ernaria solani*, *Aspergillus flavus*, *Botryosphaeria ribis*, *Colletotrich*um *gloeosporioides*, *Fusarium oxysporum*, *Helminthosporium maydis*, and *Phomopsis gossypii* (Maksimov et al., 2011). Biocontrol studies have demonstrated that the presence of lipopeptides from *B. subtilis* and *B. amyloliquefaciens* led to swelling on the nibs of developing *Sclerotinia sclerotiorum hyphae*, *Botrytis cinerea*, *Mucor* sp., *R. solani*, and *P. ultimum* (Cossus et al., 2021). Another example of such bioactivity was reported by Villa-Rodriguez et al. (2021), who observed the secretion of antibiotics by *B. cabrialesii* strain TE3^T, including fengycins, bacillaene, surfactin, rhizocticin A, bacilysin, subtilosin A, bacillibactin, and pipastatin, which caused cellular damage in *Bipolaris sorokiniana*, the causal agent of spot blotch in wheat plants.

Thus, biocontrol activity is directly related to the ability of PGPB/BCA to produce one or more of these antibiotics, where the expression of antibiotic biosynthesis genes is often dependent upon the nutritional environment provided by the host plant (Glick, 2020). Therefore, by inhibiting the growth of other soil microorganisms, these bacteria effectively create a niche for themselves to grow, function, and stimulate plant growth directly as well as indirectly.

5.2.2 Lytic enzyme production

Biological control enzymes are produced by PGPB/BCA in response to a phytopathogen attack, resulting in the elimination of phytopathogens and subsequently indirectly favoring plant growth and survival (Chaudhari Ami & Patel, 2021). Enzymes are classified into six classes by the International Union of Biochemistry (IUB): oxidoreductase, transferase, hydrolase, lyase, isomerase, and ligase, and microorganisms can generate enzymes from all six groups (Concu & Cordeiro, 2019). However, the production of enzymes involved in the degradation of the cell wall of phytopathogenic agents is one of the most widely studied biological control mechanisms, particularly against fungal pathogens (Villarreal-Delgado et al., 2018). Among the most studied lytic enzymes associated with biocontrol are chitinase and β-1,3-glucanase, due to their ability to degrade the main polysaccharides that make up the cell wall of the fungi by hydrolyzing their glycosidic bonds.

Chitinase is a hydrolytic enzyme that can degrade the chitin present in pathogens' cell walls, which compromises ~ 10–20% such as insects, fungi, and insect larvae. It is also naturally produced by a diverse range of organisms, including fungi, bacteria, yeasts, plants, actinomycetes, arthropods, and humans (Khare et al., 2018; Villarreal-Delgado et al., 2018). Endo and exochitinases are the two types of chitinases and present different activities. Endochitinase cleaves random internal points along the length of the chitinase molecule, producing diacetyl-chitobiose dimers and N-acetyl glucosamine multimers such as chitotriose and chitotetraose. Exochitinases are classified into two types: chitobiosidases, which create diacetyl chitobiose by cleaving the nonreducing ends of chitin in a stepwise way, and 1,4-glucosaminidases, which convert oligomers produced by endochitinases into monomers of N-acetyl glucosamine (Chaudhari Ami & Patel, 2021). Chitin is consumed as an energy source by bacteria; however, bacterial chitinases have been also reported as biological control agents against a range of phytopathogenic fungi that cause plant diseases, by the production of lytic enzymes.

On the other hand, β-1,3-glucanases are glycoside hydrolases found in plants, fungi, and bacteria, that cleave long chains of -1,3-glucan. The role of β-1,3-glucanase in the biological control of soilborne plant pathogens is also being investigated, and β-1,3-glucanase or other glucanase-producing microorganisms are now being used as efficient biological control agents due to their ability to modify fungal cell walls (Chaudhari Ami & Patel, 2021). Similarly, several studies have reported β-glucanases as important biological control agents, whereas Aktuganov et al. (2007) reported that β-1,3-glucanases are the main lytic enzymes involved in the in vitro control of *Bipolaris sorokiniana* by *Bacillus* sp. 739. In another study with *B. amyloliquefaciens* MET0908, the lytic activity of β-1,3-glucanases on the *Colletotrichum lagenarium* hyphae was observed using scanning electron microscopy.

5.2.3 Niche competition

In addition to mechanisms where a BCA produces a substance that kills or inhibits the functioning of the pathogen, some BCA outcompete phytopathogens for nutrients and suitable niches on the root surface, limiting plant disease incidence and severity (Glick, 2020). Several factors affect the niche presence, including plant–microbe and microbe–microbe interactions, soil characteristics, biogeography, and the favorability of the environment (Ahmed et al., 2021). Such biological activities related to niche and nutrient competition include the consumption of leached exudates, the production of siderophores, and physical niche occupation through the production of phytohormones and molecular patterns (Thakur et al., 2021; Villarreal-Delgado et al., 2018). In this manner, the growth and reproduction rate of the microorganisms plays an important role in niche competition, along with other bioactivities developed by both beneficial and pathogenic microorganisms.

Siderophore production is among the most studied mechanisms of niche competition. The production of siderophores serves as biocontrol of pathogens, as they sequester iron (Fe) and inhibit or slow down their growth and metabolic activity (Di Francesco & Baraldi, 2021). Furthermore, microbial siderophores can be reduced to donate Fe to the transport system of a plant or chelate Fe from soils. They can then engage in a ligand exchange with phytosiderophores, providing plants with this essential element to enhance their growth. Siderophores are grouped into three main families based on their functional groups: hydroxamates, catecholates, and carboxylates (Beneduzi et al., 2012). Currently, more than 500 siderophores produced by various plants, fungi, or microorganisms under Fe-limited conditions have been identified (Kügler et al., 2020).

5.2.4 Endotoxin production

Endotoxins are parasporal protein bodies composed of polypeptide units of different molecular weights, ranging from 27 to 140 kDa, and the ones produced by *Bacillus thuringiensis* are of particular interest (Villarreal-Delgado et al., 2018). To date, there are 700 reported holotypes of *B. thuringiensis* toxins, classified into 74 Crystal (Cry) and 3 Cytolytic (Cyt) groups (Paul & Das, 2021). During the sporulation phase of *B. thuringiensis*, two types of Cry and Cyt parasporal toxins are produced, where, Cry toxin is specifically toxic to the majority of insect orders including Lepidoptera, Diptera, Coleoptera, Hymenoptera, Hemiptera, Isoptera, Orthoptera, Siphonoptera, and Thisanoptera (Schünemann et al., 2014). These toxins disrupt insect cell structures, inducing osmotic cell lysis that causes significant ion leakage and loss of functional integrity (Koskey et al., 2021).

B. thuringiensis strains and their insecticidal crystalline proteins (ICPs) account for 90%–95% of the bioinsecticides used throughout the world, with over 180 *B. thuringensis* products registered in the US Environmental Protection Agency, approximately 276 *B. thuringensis* microbial formulations registered in China (Paul & Das, 2021), and 76 registered in Mexico. Furthermore, some *B. thuringiensis*, and *B. cereus* strains produce a third group of vegetative insecticidal proteins (Vips), including Vip1, Vip2, and Vip3, where Vip1 and Vip2, which are binary toxins that have coleopteran specificity, whereas Vip3 toxins have lepidopteran specificity. These proteins induce insect gut paralysis and complete lysis of gut epithelium, and are synthesized during the vegetative growth phase of the bacterium (Paul & Das, 2021). Among the different Vips, the well-characterized Vip3A induces lethal toxicity against larvae of *Agrotis ipsilon* and *Spodoptera frugiperda* (Xianmei Yu et al., 2011), while Vip3Aa14 against larvae of *Spodoptera litura* and *Plutella xylostella* (Bhalla et al., 2005).

5.2.5 Mycoparasitism

Endotoxins play a crucial role in fungal cell wall degradation, which is necessary for mycoparasitism to take place. Mycoparasitism is the direct attack of one fungus by another fungus that exploits the targeted fungi as a nutrient source (Moreno-Ruiz et al., 2020). The events leading to mycoparasitism are complex and take place as follows: chemotropic growth, host recognition, coiling and appressoria formation, secretion of hydrolytic enzymes, penetrations of the hyphae, and lysis of the host. These processes also include sequential events, involving the cycle of recognition by the binding of carbohydrates in the fungal cell wall to lectins on the target fungus, followed by hyphal coiling and appressoria formation, which contains a higher amount of osmotic solutes such as glycerol, and induces penetration. The attacking fungus produces several cell-wall-degrading enzymes, including glucanases, chitinases, and proteases, and the cumulative action of these compounds results in parasitism of the target fungus and dissolution of the cell walls. Appressoria formation holes can be produced in the target fungus, allowing the direct entry of the attacking fungi hyphae into the host lumen and, ultimately, the killing of the host (Singh et al., 2018). Genera of the order Hypocreales (Sordariomycetes, Ascomycota) such as *Trichoderma*, *Metarhizium,* and *Beauveria* are potent mycoparasites that target plant-pathogenic fungi and insect pests that cause severe damage to crops each year (Wang et al., 2021).

5.3 Mechanisms with dual effect

Several activities can be classified under both mechanisms, for example, the production of siderophores, the production of volatile compounds, and ACC deaminase activity (Fig. 5.3) (Santoyo et al., 2019). Also, PGPB can exhibit one or more growth promotion mechanisms and may occur multiple times at different stages of the plant growth cycle (Glick, 2012).

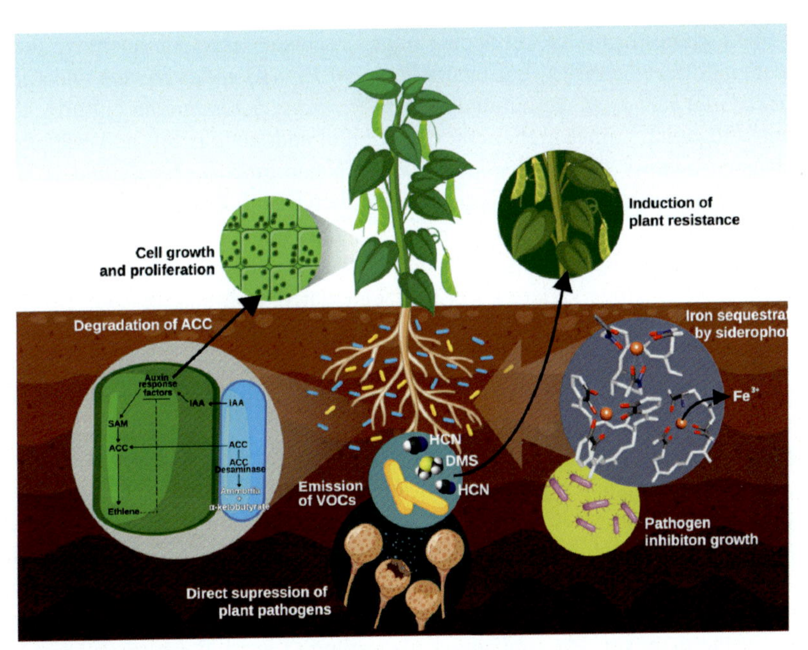

FIGURE 5.3 Mechanisms with the dual effect of PGPB. Plant growth promoting bacteria can carry out beneficial effects on plants through mechanisms with dual effects, providing nutrients to the plant and inhibiting phytopathogens, such as the production of some phytohormones, quorum sensing, or volatile organic compounds (VOCs).

5.3.1 Phytohormones production

A large number of PGPB present in the rhizosphere can regulate plant growth through the production of secondary metabolites, especially phytohormones, which mediate nutrient mobilization, alter plant structures, and trigger biochemical pathways that lead to overall plant improvement, as reflected in yield and prevalence (Etesami & Maheshwari, 2018).

Abscisic acid is a drought stress mitigation agent produced when plant roots detect the lack of water in the environment using transmembrane proteins such as histidine kinase and aquaporins. Once these phytohormones are produced, they are transported from the roots to the aerial part of the plant, where they can mediate stomatal closure to avoid water loss through transpiration. In addition, they promote the branching of the roots to extend their coverage and thus obtain water resources (Yadav, 2020; Cohen et al., 2009).

Salicylic acid (SA) is a phytohormone with functions that involve the activation and regulation of responses to biotic and abiotic stress, highlighting its role as a defense against pathogenic microorganisms. SA tends to accumulate in infected areas to later spread throughout the rest of the plant, inducing systemic acquired resistance in the healthy parts of the plant (Koo et al., 2020; Maruri-López et al., 2019).

Field studies using SA-producing microorganisms have demonstrated their beneficial effect on crops. For example, when *Pseudomonas tremae* and *Curtobacterium herbarum* were inoculated in *Nicotiana benthamiana* cultures, the plants showed 2.3 times higher concentration of endogenous SA and exhibited resistance against *P. syringae* pv. *tabaci*. In addition, *N. benthamiana* and *N. tabacum* presented fewer lesions when infected with *P. syringae* pv. *tabaci*, also exhibiting an improvement in the growth of height and weight (Islam et al., 2020). In corn cultivation, the seeds of the drought-tolerant variety SWL-2002 and drought-sensitive variety CZP-2001 inoculated with a suspension of *Planomicrobium chinense* and *Bacillus cereus* demonstrated a reduction of lipid peroxidation in 33% and 23%, respectively. In addition, the content of proline and the activities of antioxidant enzymes decreased by 32% and 38%, respectively. The inoculation also increased the content of Ca, Mg, K Cu, Co, Fe, and Zn in plant shoots and rhizosphere by more than 52%. Moreover, there was an increase in soil organic matter and total nitrogen, and carbon/nitrogen ratio by 42% (Khan et al., 2020). In sunflower crops, the inoculation of *Pseudomonas aeruginosa* PF23EPS and its mutant PF23EPS (polysaccharide excretion-deficient strain), along with the use of SA extract, resulted in a significant improvement in plant growth under saline conditions and in the presence of the pathogen *Macrophomina phaseolina* (Tewari & Arora, 2018).

Jasmonic acid has been extensively studied due to its ability to mitigate biotic stress in plants and mediate crop responses against pests and necrotrophic pathogens (Veselova et al., 2015). PGPB producers of this phytohormone can regulate the induced systemic response in plants, enabling them to defend themselves on time against phytopathogens. This mechanism triggers an increase in the production activities of enzymes such as chitinase, peroxidase, polyphenol oxidase, and phenylalanine ammonia-lyase, in addition to increasing the production of phytoalexin (Verhagen et al., 2004; Veselova et al., 2015).

5.3.2 Siderophores production

Iron (Fe) is an essential element for different biological processes such as oxygen metabolism, DNA and RNA synthesis, electron transfer, and enzymatic processes. Despite being the fourth most abundant element on Earth, it is commonly found in insoluble compounds in the presence of oxygen, such as hydroxides and oxyhydroxides, which are not readily useable by plants (Etesami & Maheshwari, 2018). When faced with a deficit of this nutrient, the microorganisms produce chelating compounds known as siderophores (Yadav, 2020). Siderophores are low molecular weight molecules that capture and chelate the insoluble Fe^{3+}, making it soluble and useable Fe^{2+} available for both microorganisms and plants. The siderophore-Fe complexes formed are taken up by bacterial and plant cells, where the complex is subsequently decomposed. Siderophores are excellent agents for obtaining Fe, and converting microorganisms possessing this mechanism as promising PGPB, within which the genus *Pseudomonas* has the highest yield of siderophore secretion (Yadav et al., 2020; Etesami & Maheshwari, 2018; Rajkumar et al., 2010).

Siderophores are also known to have an affinity for other heavy metals (Rajkumar et al., 2010; Xiumei Yu et al., 2014). However, due to the high abundance of iron in the soil, the solubilization rate of Fe is significantly higher compared to other related metals, so this does not pose a risk to the effectiveness of this mechanism (Cohen et al., 2009; Etesami & Maheshwari, 2018).

Studies on microorganism–plant interactions that regulate the siderophore production have shown that secretion of these bacterial chelating compounds is a crucial process for obtaining Fe in calcareous soils, as well as in other arable soils. This highlights the importance of PGPB as biofertilizers, which can help to reduce the use of phosphorous agrochemicals (Hider & Kong, 2010; Radzki et al., 2013; Thijs et al., 2016). On the other hand, it has been demonstrated that direct promotion of growth by using radioactively labeled siderophores as the only source of Fe, results in a higher level of chlorophyll content in plants in the presence of siderophores compared to those not treated with siderophores (Gupta et al., 2015).

5.3.3 Quorum sensing

Some PGPM, particularly bacteria, can recognize individuals of the same species and other agents present in the environment. This recognition is achieved through chemical signaling, allowing them to census the population and presence of other microorganisms; this function is known as quorum sensing. When bacteria reproduce and reach a certain cell density, they release signaling metabolites that trigger biochemical pathways that allow them to recognize each other and carry out cooperative mechanisms (Cornforth et al., 2014; Olanrewaju et al., 2017).

Signaling molecules, commonly referred to as auto-signaling molecules, increase in concentration as the bacterial population grows; this rise in metabolic activities reaches thresholds above which the bacterial population can act as a single biological entity (Olanrewaju et al., 2017). These auto-indicators vary according to the microorganism; in Gram-negative bacteria, N-acyl homoserine lactones are the main regulators of quorum sensing, while for Gram-positive bacteria, incomplete cyclic peptides, AI-2, and butyrolactone control the cellular concentration sensing functions in the medium (Singh et al., 2017).

The quorum sensing mechanism bring beneficial effects such as biofilm production, bioluminescence, nutrient mobilization, and bioremediation functions. However, pathogens with the same cellular recognition capacity can activate detrimental mechanisms such as toxin production or a significant increase in its virulence (Olanrewaju et al., 2017). Therefore, from the perspective of promoting plant development, quorum sensing functions is a promising mechanism for PGPM that possess them. In this manner, the disruption of the recognition of pathogens represents two approaches: promoting plant growth and enhance biocontrol (Yadav, 2020).

There are several mechanisms by which PGPM can stop quorum sensing in pathogens. Inhibiting quorum sensing functions among harmful microorganisms can stop disease proliferation and prevent plant growth inhibition. One example is the secretion of the enzyme lactonase, which can degrade the key autoinducer of signaling among pathogens (Olanrewaju et al., 2017).

Part II

Development of next-generation bacterial inoculants

Chapter 6

Bioprospection of beneficial bacteria for bioproducts innovation in agriculture

Alondra María Díaz-Rodríguez[1], **Roel Alejandro Chávez-Luzanía**[1], **Edith Rojas Anaya**[2] **and Sergio de los Santos Villalobos**[1]

[1]*Instituto Tecnológico de Sonora, Ciudad Obregón, Sonora, Mexico;* [2]*Centro Nacional de Recursos Genéticos, INIFAP, Tepatitlán de Morelos, Jalisco, Mexico*

The production of microbial inoculants is a complex process, from the strain selection to its application in the field. This process generally involves the selection of strains of beneficial microorganisms, their cultivation in specific growth media, and their multiplication in large quantities for extensive application in the field. This may vary depending on the type of microorganism used (Fig. 6.1). For the selection of microorganisms, challenges such as the ecological viability in the soil, the integration of plant growth-promoting bacteria (PGPB), and the aptitude of the microorganisms to carry out the promotion or biological control mechanisms expected under in situ conditions have to be considered (Valenzuela-Ruiz et al., 2018). The selection of the microorganisms to be inoculated is very critical. The main features of PGPB strains used in bioformulations must: (1) Be nonpathogenic to the *Animalia* and *Plantae* kingdoms (Ferreira et al., 2019), (2) Grow in an artificial medium and survive in carriers (Herrmann & Lesueur, 2013), (3) Able to survive, compete with other microbiota, (4) Colonize the root surface and plant tissues (Di Benedetto et al., 2017), (5) Be compatible with crop production practices (Parnell et al., 2016), and (6) Be stress tolerant (Valenzuela-Aragon et al., 2019).

There are many studies on the choice of the best microorganism candidates, their isolation, and characterization from the rhizosphere of various crops. The main approach is through culture-dependent methods based on isolation, characterization, identification, and selection according to its potential use. This involves a lot of research and development time since it has to be validated in cultivation. The aim is to have strains with desired characteristics and properties; in these methods, the quantitative distribution (relative

New Insights, Trends, and Challenges in the Development and Applications of Microbial Inoculants in Agriculture
https://doi.org/10.1016/B978-0-443-18855-8.00006-0

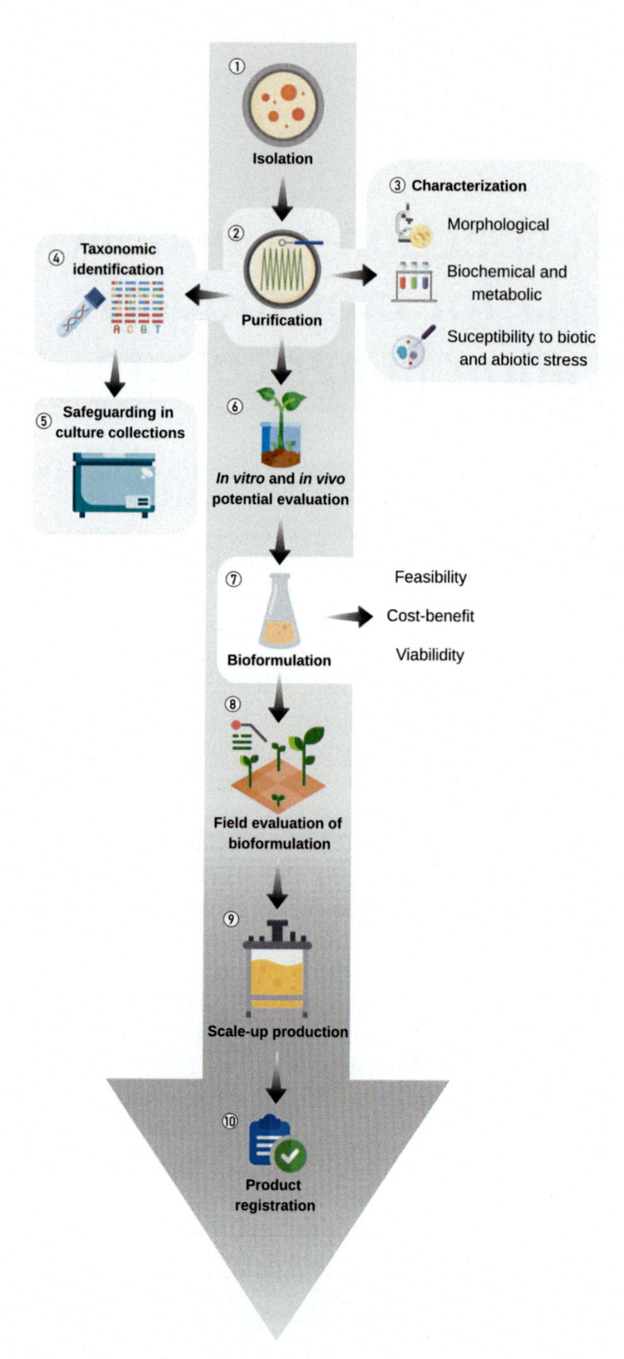

FIGURE 6.1 Strategy for the development of a microbial inoculant. The process of developing microbial inoculants involves several steps: First, is the (1) screening of microorganisms isolated

abundance) is not considered, but the technological robustness of the strains is evaluated (Di Benedetto et al., 2019).

The abundance of the inoculum after its application is an important point to take into account, although the population of the microorganism in the applied matrix does not always mean that it is affecting the host plant. However, multiple reports have shown positive correlations between the presence of the inoculum and the beneficial effects of growth promotion and biocontrol on the plant (Atkins et al., 2005; Baudoin et al., 2010; Couillerot et al., 2010; Mendis et al., 2018; Stets et al., 2015; Zhang et al., 2020).

Several techniques exist to detect microbes under field conditions to investigate the correlation between the introduction of microorganisms and the resulting impact on the crop. However, due to the intricate nature of agricultural soil and rhizosphere, the identification of specific microorganisms becomes challenging. This is due to the diversity of the microbiome and the different components within the matrix, which can potentially hinder downstream analysis procedures (Rilling et al., 2019). The main methods for this purpose include (1) β-Galactosidase (*lac*Z) reporter gene, (2) β-Glucuronidase (*gus*A) reporter gene, (3) Bacterial luciferase (lux), (4) Green fluorescent protein (GFP), (5) Enzyme-linked immunosorbent assay (ELISA), (6) Immunofluorescence, (7) Immunogold, (8) Immunoblot, (9) Fluorescent in situ hybridization (FISH), (10) polymerase chain reaction most probable number (NMP-PCR), (11) Competitive PCR (C-PCR), and (12) Quantitative PCR (q-PCR) (Bustin & Huggett, 2017; Navarro et al., 2015; Rilling et al., 2019).

The monitoring of the inoculum population in the matrix of interest then becomes a tool for the design of microbial inoculants, and a basis for achieving the maximum benefits for the crops also contributes to the understanding of the interaction between microorganisms and the plants, and how they can, or cannot, persist under field conditions (Chávez-Luzanía et al., 2022).

It is determinant to isolate, characterize (gram staining, microscope observation, macroscopic characterization, phenotypic test, and taxonomic affiliation), and identify promising PGPB through metabolic tests to seek capabilities of plant growth promotion and biocontrol against phytopathogens.

from the soil, rhizosphere, and/or plant tissues, followed by (2) the isolation of individual bacterial strains through techniques such as dilution plating or selective media. Then (3) different analyses are conducted to evaluate their capabilities (selection filters), such as metabolic tests, stress susceptibility assessments, and tests with native soil microbiota. The hemolytic analysis is also necessary to exclude pathogenic microorganisms. (4) The selected strains for the microbial inoculant must be taxonomically affiliated and (5) conserved in certified culture collections. (6) *In vitro* and *in vivo* analyses are then carried out to determine the positive effect on plants, followed by (7) inoculant production based on the nutritional requirements of the selected strains and the choice of the most suitable carrier. (8) Then, the inoculant is further evaluated in field trials. (9) Successful results may lead to the scale-up of the product and finally (10) it can be registered for commercial use.

It is also essential to perform biosafety analysis to discard potential pathogenic microorganisms toward humans, plants, and/or animals, such as hemolytic test (Fig. 6.1(1–5)) (Di Benedetto et al., 2019; Villa-Rodríguez et al., 2019). The choice of strains for bioformulation depends mainly on the objectives of the bioinoculant formulation, that is, the desired effects to be obtained, as well as the crop to be applied (Stamenković et al., 2018).

Recent works have studied the benefits of using native bacteria as inoculants to enhance plant growth, suggesting that the strains are already adapted to the local agroecosystem, increasing the survival chances of the inoculum and potentially conferring positive effect on the development of the plants under previously presented stresses (Banerjee et al., 2017; Z. Qiu et al., 2019). In addition to direct isolation from bulk soil or rhizosphere, the selection of potential PGPB can be performed by plant-assisted selection, where plants select which strains to associate with from the inoculation of an introduced bacterial population. This strategy could be used under different host plants and abiotic/biotic conditions (Valenzuela-Aragon et al., 2019).

Some commercial inoculants depend on the inoculation of a single strain, limiting us to a smaller diversity of plant growth promotion or biological control mechanisms. To overcome this problem, microbial consortia products or the coinoculation of different PGPB-based bioformulations allow the combination of different potential mechanisms without the need to resort to genetic engineering, proving to be a better approach for the growth and development of a greater diversity of crops (Odoh et al., 2020; Trabelsi & Mhamdi, 2013). Thilagar et al. (2016) showed that the inoculation of a microbial consortium (AM fungus *F. mosseae* + *B. sonorensis*) with half the recommended dose of chemical fertilization can achieve the same yield values in chili as with only conventional dosage. The application of a consortium-based biofertilizer, containing strains of *P. fluorescens* 1N (potential N_2 fixation), *B. subtilis* B9, and *B. amyloliquafaciens* E19 (ability to break down protein, cellulose, and starch) and a soil yeast, *Candida tropicalis* HY (phosphate solubilizer and promote root systems) reported an increase in plant biomass, N uptake, and maturity grain yield in field trials of rice crops (Nguyen et al., 2017). Microbial consortia have been also evaluated as biological control agents (BCA) with an expectation of synergistic interactions among BCA. A theoretical model proposed by Xu & Jeger (2013) suggests that synergy may result from the combined use of two BCA with different biocontrol mechanisms, under spatially heterogeneous conditions. Likewise, experimental studies have found that combinations of BCA have, at least, similar efficacies to the most efficacious one used alone; Silva et al. (2018) tested the efficacy of five bacterial isolates (*Brevibacterium* sp., *Bacillus* spp. *Paenibacillus* sp. and *Serratia* sp.) and their dual combinations to control bole rot of sisal in vitro and in field experiments, where they found that at least one of the bacterial combinations reduced significantly the disease incidence levels. Similar results were obtained when using consortia of three or more strains; Srinivasan & Mathivanan (2009) used consortia of three (*Bacillus* sp., *B. licheniformis*, and *Pseudomonas aeruginosa*) and four (*Bacillus* sp., *B. licheniformis*, *P. aeruginosa*, and *Streptomyces fradiae*) different bacterial strains to control sunflower necrosis virus disease (SNVD) and showed to reduce the disease up to 51.4%, compared to control.

Other important characteristics to consider in the selection of strains are their effects under the biotic and abiotic variables of agrosystems, that is, their response to pH, salinity, moisture content, and climatic fluctuations, the interactions and competitiveness with native microbiota, the ability to persist in the soil, plants and/or seeds and to produce the growth promotion mechanisms on the target crops, their genetic stability, and the tolerance to environmental constraints (i.e., salt, hydric stress, temperature, and even pesticides concentrations) (Díaz-Rodríguez et al., 2019; Ibarra-Villarreal et al., 2021; Nadeem et al., 2014). For this, it is necessary to conduct in vitro and in vivo experiments under different conditions (Fig. 6.1(6)); it is recommended that trials these include the study of colonization and establishment in soil and root (Romano et al., 2020), measurement of effects on plants (Robles-Montoya et al., 2020), plant–microbe interactions at the transcriptomic level (see Chapter 7) (Moradi et al., 2021), and effects on native soil microbiota (Ye et al., 2019).

Some studies have indicated that native PGPB can enhance plant growth in the presence of pesticides (Mishra & Sundari, 2015; Verma et al., 2016), and these had no significant impact on the present microbiota, instead, their use increased the abundance of native microbial groups with biocontrol activity, while the chemical pesticides and fertilizers cause a reduction of microbial community richness and diversity (Escribano-Viana et al., 2018; Xiong et al., 2017). The use of bacteria with the ability to promote plant growth or biocontrol can be used in conjunction with lower pesticide applications stimulating increased production and helping preserve the beneficial soil microbiota (Peláez-Álvarez et al., 2016).

Biological control, using specific BCA strains, has been demonstrated against many plant pathogens, through various mechanisms, such as antagonisms (parasitism and antibiosis), and competition for space and nutrients, that is, *Burkholderia cepacia* XXVI (siderophore production) (de los Santos-Villalobos et al., 2012), *Bacillus paralicheniformis* (lipopeptides and antibiotic biosynthesis) (Valenzuela-Ruiz et al., 2019), and *Streptomycetes* (inducing the expression of plant defense pathways) (Vurukonda et al., 2018). However, as previously mentioned, the success of this type of alternative in the field is variable since its efficiency is altered by stimuli from the environment that surrounds them. Thus, tests of a minimum inhibitory concentration of pesticides on biological control agents must be developed to know the minimum concentration of pesticides that in coinoculation with biological control agents do not interfere with each other's development, but contribute to the dual control of phytopathogens (Peláez-Álvarez et al., 2016; Verma et al., 2016).

Once the microorganisms to be produced have been defined, they are multiplied in an optimal culture medium with the appropriate growth conditions; for this purpose, screening is carried out to select the optimal medium and parameters for the fermentation process. The objective of the screening is to minimize fermentation costs and produce greater quantities of the microorganism. Compounds that function as carriers are added to the bioproduct in this step to maintain the microbial biomass obtained from fermentation in good conditions and to be easily applied in the field (Fig. 6.1(7)) (Kamilova et al., 2015). Bioinoculants can be formulated based on liquid or solid carriers (see Chapter 11) (Alori & Babalola, 2018). It is important to conduct field trials considering the agroclimatic conditions of the crops of interest and routine field practices (seed treatment, machinery, and common use of agrochemicals) to evaluate the final bioformulation in the establishment, growth promotion, and/or biocontrol of phytopathogens and yield performance (Fig. 6.1(8)) (Bashan et al., 2014). Finally, if all of the above steps had the expected and consistent results, it is possible to move on to scalling up using bioreactors of different volumes (Fig. 6.1(9)) (Kamilova et al., 2015).

Chapter 7

Molecular and biochemical responses of plants to bacterial inoculants

Luis Abraham Chaparro-Encinas[1], Fannie Isela Parra Cota[2] and Sergio de los Santos Villalobos[3]

[1]*Universidad Autónoma Agraria Antonio Narro, Torreón, Coahuila, Mexico;* [2]*Campo Experimental Norman E. Borlaug, INIFAP, Ciudad Obregón, Sonora, Mexico;* [3]*Instituto Tecnológico de Sonora, Ciudad Obregón, Sonora, Mexico*

Research on plant growth-promoting bacteria (PGPB) has been on the rise since the term was first used in the 1970s and was promoted by the characterization of nitrogen-fixing bacteria and how they influence crop yield (Vessey, 2003). Since then, a plethora of plant growth-promoting mechanisms have been discovered, and it is projected that there is much more to be found. The study of plant—microorganism interactions involves the qualification and quantification of molecules at different levels. These estimations have been improved with the advent of omics sciences and advances in spectrometry.

Microbial communities are extremely diverse, most of them inhabit the rhizosphere and interact through chemical communication; in turn, plants exert their influence on the microbiota through primary and secondary metabolites called exudates. Plants can select the structure of their microbiome through exudates, as they can reconstitute the microbiome at different stages of development, genotypes, and domestication (Schlaeppi et al., 2014). Exudates constitute up to 50% of photosynthesis carbon production and consist of molecules of miscellaneous types, which can be of protein and nonprotein nature, signaling or nutrient, and antagonistic or non-antagonistic to microorganisms (Baetz & Martinoia, 2014). Advances in untargeted metabolomics have identified hundreds of metabolites in exudates from model plants such as *Arabidopsis thaliana* (Strehmel et al., 2014). These metabolites accomplish diverse roles; for example, phenolic compounds inhibit germination of other plants, volatile organic compounds (VOCs) or terpenes are exuded by insect attack to attract entomopathogenic nematodes, and strigolactones in the root exudates induce hyphal branching in arbuscular mycorrhizal fungi, which is

New Insights, Trends, and Challenges in the Development and Applications of Microbial Inoculants in Agriculture
https://doi.org/10.1016/B978-0-443-18855-8.00007-2

required for colonization (De Coninck et al., 2015). In turn, it has been observed that plants promote the colonization of PGPB through the synthesis of polysaccharides, surfactants, and the formation of biofilms. This response is perceived by the bacteria that occupy this niche, emitting phytohormones, solubilizing nutrients, and presenting antagonism with pathogenic organisms (Zhang et al., 2016).

Metabolic patterns in plants are environmental-responsive, employing a vast collection of mechanisms, particularly secondary metabolites. The spatial-temporal synthesis of secondary metabolites potentially mitigates the impact of abiotic and biotic stress, most of which depends on recruiting of beneficial bacteria and mycorrhizae (Hong et al., 2022) (Fig. 7.1).

It has been observed in maize that during nitrogen starvation, an increase in flavone synthesis results in the enrichment of Oxalobacteraceae, induction of lateral root growth, and consequently, enhancement of nitrogen uptake (Yu et al., 2021).

Coumarins are other regulators to remodel the structure of the microbiota under biotic and abiotic stress. Fraxetin acts as a recruiter of iron (III)-reducing bacteria under iron deprivation; such bacteria convert unavailable ferric iron into soluble ferrous compounds (Harbort et al., 2020). On the other hand, scopoletin selectively inhibits the fungal pathogen *Fusarium oxysporum* and *Verticillium dahliae* while inducing colonization of beneficial bacteria *Pseudomonas simiae* WCS417 and *P. capeferrum* WCS358 in *Arabidopsis thaliana* (Stringlis et al., 2018). Notable examples of pathogen inhibition and beneficial bacteria recruiting were observed through the synthesis of camalexin (Koprivova et al., 2019) and benzoxazinoids (Kudjordjie et al., 2019).

Secondary metabolites also exert a strong influence on fungal populations. For example, several defense compounds, such as indole glucosinolates, can control the pathogenicity of *Colletotricum tofieldiae* (Hiruma et al., 2016). Also, it has been reported that strigolactones promote the growth of arbuscular

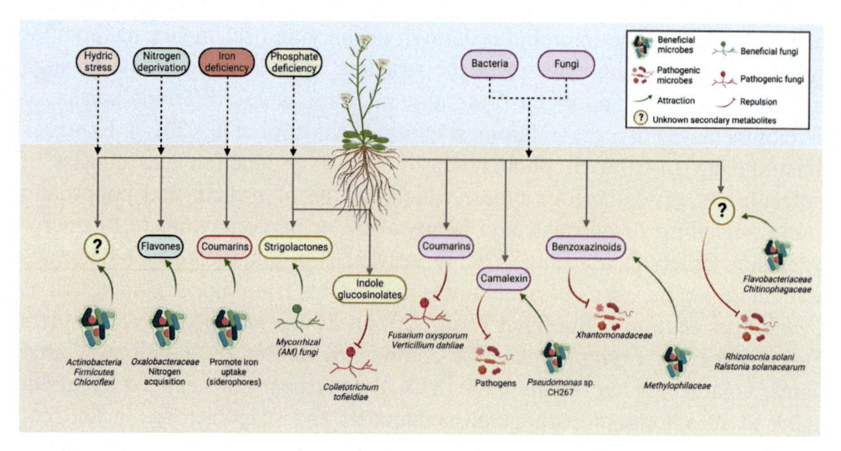

FIGURE 7.1 The metabolic response of plants under abiotic and biotic stress changes the structure of the microbiome by recruiting beneficial bacteria and fungi while inhibiting pathogenic microbes (Hong et al., 2022).

mycorrhizal fungi, providing positive effects under phosphate starvation (Kapulnik & Koltai, 2016).

These metabolic responses have a complex background of expression, silencing, and regulation mechanisms at the genomic, transcriptional, and epigenetic levels. These mechanisms are activated by a signaling system of intracellular and extracellular molecular receptors that recognize pathogen-associated molecular patterns (PAMPs), microbe-associated molecular patterns (MAMPs), phytohormones, and other inducers such as VOCs and siderophores (Rusnac & Zheng, 2018). The perception of these patterns induces activities that affect not only the contacted tissue but the entire plant. This type of response is called systemic resistance, which is physiologically defined as an improved state of defensive capacity that protects tissues not exposed to the biotic factor and can be induced by herbivorous insects, infection by pathogens, and colonization of PGPM/PGPB (Ádám et al., 2018) (Fig. 7.2).

The two forms of systemic resistance are systemic acquired resistance (SAR), which is activated by plant pathogens, and induced systemic resistance (ISR), which is promoted by beneficial microorganisms (Choudhary & Johri, 2009). SAR is characterized by increasing salicylic acid (SA) concentration and influencing pathogenesis-related (PR) gene expression. At the same time, SAR involves independent pathways of SA, which can vary depending on the microorganism, and both ethylene (ET) and jasmonic acid (JA) play vital roles in the regulation of PR and defensin (PDF) genes (Romera et al., 2019). Furthermore, there are damage-associated molecular patterns (DAMPs) composed of the resulting products of plant degradation by pathogens and herbivores (Hou et al., 2019). Due to the high phylogenomic similarity between pathogens and plant growth-promoting microorganisms (PGPM), the line on which plants distinguish pathogens and mutualistic organisms is not fully defined.

The microbiota recognition system in ISR and SAR begins with the identification of MAMP/PAMP/DAMP through pattern recognition receptors (PRR) (Schwessinger & Ronald, 2012). These receptors are part of pattern-activated immunity (PTI) and involve activation of the MAPK (mitogen-activated protein kinases) cascade, Ca^{2+} signaling, transcriptional reprogramming, and production of reactive oxygen species (ROS) (Bigeard & Hirt, 2018). Specifically, SAR involves the activation of pathogenesis-related (PR) genes through the nonexpression of PR genes1 (NPR1), which influences the transcription factors of the TGA family (TGACG binding) and the WRKY family (named after the highly conserved amino acids sequence WRKYGQK) (Phukan et al., 2016). However, some pathogens have developed virulence effectors to evade PTI activation; in response, plants that are resistant to these pathogens possess NB-LRR (nucleotide-binding leucine-rich repeat) receptors, which trigger effector-triggered immunity (ETI) (Han & Jung, 2013). This is a form of resistance gene-by-gene, which is highly specific for each pathogen and generally involves the programmed-cell death in the infected tissue, as well as the production of ROS, chitinases, and phytoalexins (Ádám et al., 2018).

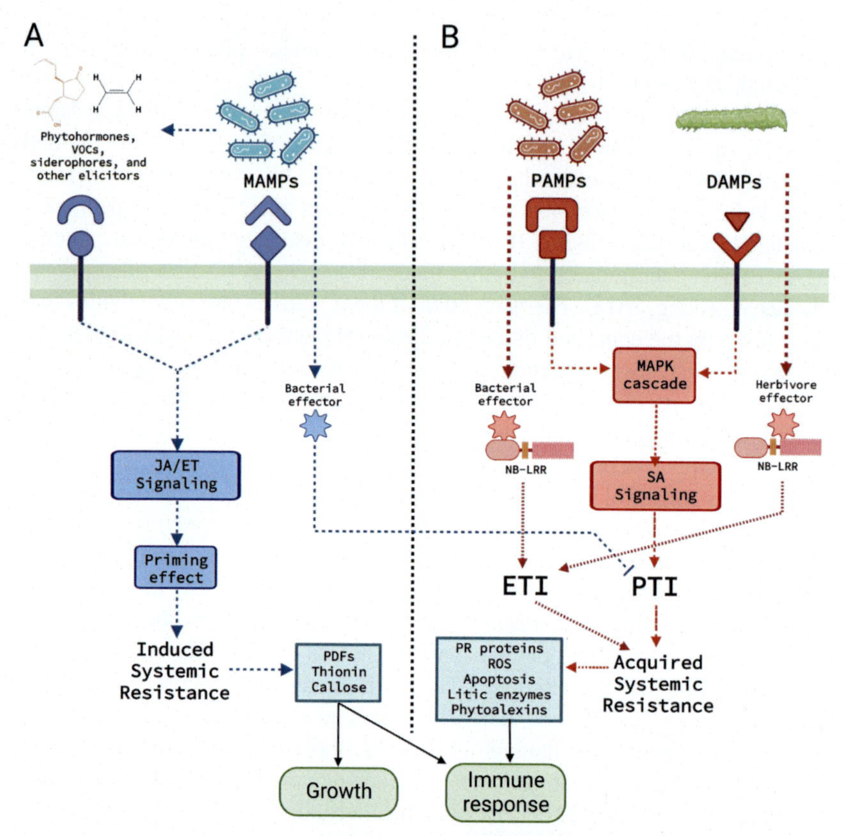

FIGURE 7.2 Plant recognition mechanisms of MAMP/PAMP/DAMP. (A) Beneficial interactions for the host plant. (B) Pathogenic interactions for the host plant. *DAMPs*, damage-associated molecular patterns; *ET*, ethylene; *ETI*, effector-triggered immunity; *JA*, jasmonic acid; *MAMPs*, microbe-associated molecular patterns; *NB-LRR*, nucleotide-binding leucine-rich repeated receptor; *PAMPs*, pathogen-associated molecular patterns; *PDFs*, defensins; *PR*, pathogenesis-related proteins; *PTI*, pattern-triggered immunity; *ROS*, reactive oxygen species; *SA*, salicylic acid (Chaparro-Encinas et al., 2020).

On the other hand, MAMPs that activate ISR usually (but not exclusively) proceed from beneficial rhizospheric microorganisms by elicitors such as chitin/chitosan, beta-glucans, xylanase, and oligosaccharides, among others (Depuydt et al., 2009). A wide diversity of MAMPs from PGPM has also been reported, mainly from *Bacillus* and *Pseudomonas* species, including flagellin (flg), lipopolysaccharides (LPS), elongation factors (EF-Tu), peptidoglycans, siderophores, and VOCs, among others (Newman et al., 2013). These effectors are perceived by the PRR receptors, which are classified based on the exposed regions in the apoplast. The two main types are the LRR-type receptors and the lysine motif (LysM) (Desaki et al., 2019). In addition, it is important to note that the influence of PGPM is not limited to MAMP recognition alone; it is necessary to consider the ability to activate ISR through the synthesis of exogenous hormones (auxins, cytokinins, etc.) and VOCs. Recent studies have extensively reported these factors as important activators of pathogen response in *Bacillus* strains (Arrebola et al., 2010; Ryu et al., 2004; Zhao et al., 2019).

The regulatory effect of PGPM elicitors in ISR does not directly influence defense mechanisms but rather promotes a hypersensitivity phenomenon known as priming, which triggers a defensive response throughout the plant (Wang et al., 2014). Priming is an intrinsic part of ISR as the plant takes defensive measures against potential damage while also preparing its resistance system for a faster and stronger reaction in the future (Mauch-Mani et al., 2017). This phenomenon regulates the transcription factors related to JA and ET, such as those belonging to the AP2/ERF family, which in turn promote the production of plant defensins (PDFs) and thionines, and promotes the reinforcement of cell walls through callose deposition (Bienert et al., 2014). Several PGPM and PGPB have been reported to suppress response mechanisms that limit the colonization capacity of pathogens, such as flagellin recognition (Rodriguez et al., 2019), facilitating the establishment of PGPB in the rhizosphere (Malik et al., 2020). However, it is important to note that PGPB must be at a minimal concentration to have effects on ISR. It has been proposed that such concentration should reach 10^5-10^7 colony-forming units (CFU) per gram of root and must proliferate for variable periods ranging from days to weeks to establish a lasting mutual relationship (Bakker et al., 2013; Pieterse et al., 2014). This relationship is dynamic because plants can modify or exert pressure on the population, diversity, and metabolic activity of the microbiome through rhizodeposition and the production of secondary metabolites (Pascale et al., 2020).

Another recently discovered plant–microbiota interaction mechanism involves the influenceof noncoding RNAs (ncRNAs), which can interfere with the expression of trans-specific genes. Some pathogens (*i.e.*, *Botrytis cinerea*) have been shown to emit small interfering RNAs (siRNAs) that are complementary to numerous defense genes, thereby inhibiting their expression by cross-kingdom/organism RNA interference (RNAi) (Fig. 7.3A) (Cai et al., 2018). In turn, plants can exert a phenomenon known as host-induced gene silencing (HIGS), which commonly involves the generation of long, double-stranded RNAs. These RNAs, known as lncRNA, interfere with large regions of the pathogen's genome that are related to virulence capacity, leading to complete inhibition of pathogenesis (Cai et al., 2019; Hudzik et al., 2020).

PGPB also participates in the regulation of the systemic response through small RNAs. Beneficial microorganisms (e.g., *Bacillus*, *Pseudomonas*) have been shown to suppress the expression of micro-RNA (miRNA), which are responsible for inhibiting the expression of resistance genes (Fig. 7.3B). Thus, plants inoculated with PGPM whose specific miRNA expression patterns were inhibited showed more resistance to pathogens. Such findings were reported in maize by *Bacillus velezensis* FZB42 (Xie et al., 2019) and *Arabidopsis* by *Bacillus cereus* AR156 (Jiang et al., 2020).

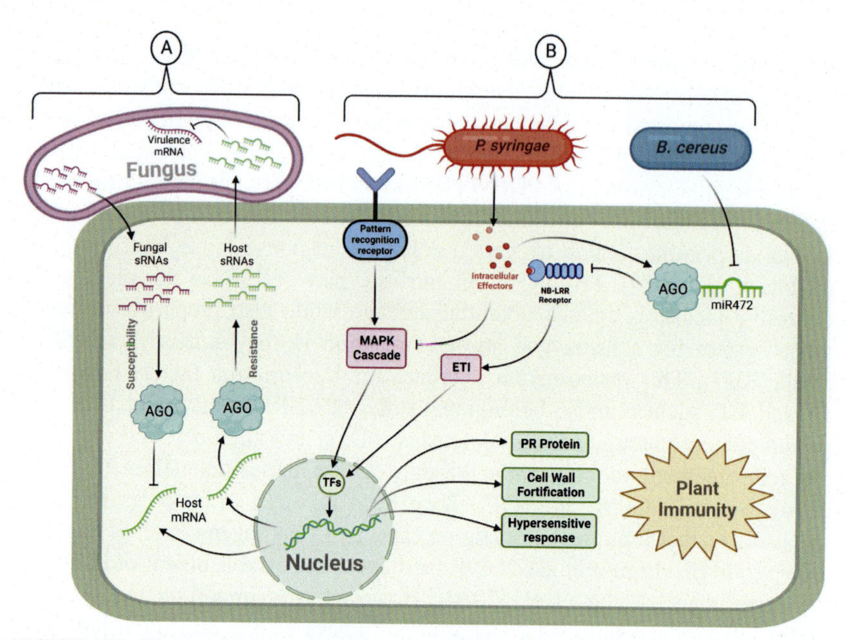

FIGURE 7.3 Plant–microbiota interactions mediated with sRNAs and their effect on susceptibility/resistance and plant immunity. (A) Gene silencing by pathogenic fungus through sRNAs/AGO complex targeting resistance transcripts, and the plant response by emitting extracellularly sRNAs to inhibit virulence transcripts in fungus to promote resistance. (B) Plant sRNA called miR472 inhibits *P. syringae* recognition by NB-LRR receptors, blocking ETI and causing disease; in turn, the presence of *B. cereus* inhibits miR472 promoting ETI and consequently plant immunity. *AGO*, argonaute protein; *ETI*, effector-triggered immunity; *NB-LRR*, nucleotide-binding leucine-rich repeated receptor; *PR protein*, pathogenesis-related protein; *TFs*, transcription factors (Cai et al., 2019; Jiang et al., 2020).

The above discussion demonstrates how plants respond at different levels to biotic factors, both pathogenic and beneficial interactions. However, many of these mechanisms remain undetermined or poorly described, presenting a window of opportunity for research. Therefore, it is advisable to adopt comprehensive approaches that combine the use of a multiomic approach with classical techniques of microbiology, immunology, and agronomy. By integrating these approaches, a more thorough understanding of plant-microorganism interactions can be achieved.

7.1 Plant's defense triggered by beneficial microorganisms: systemic-induced resistance

Plant growth promotion microorganisms, which are closely associated with the root system of plants, have the ability to prime plants against subsequent pathogen infection in systemic tissues. This type of resistance is known as induced systemic resistance (ISR) (Pieterse et al., 2014). PGPM-mediated ISR was initially thought to be similar to pathogen-induced SAR (see Section 3.4). However, research has provided evidence that induced resistance by *Pseudomonas fluorescens* WCS417r inoculation was developed without the accumulation of the PR proteins, which are characteristic for SAR (Pieterse et al., 1996). Moreover, transgenic *Arabidopsis* plants that are unable to accumulate SA, provided genetic evidence that *P. fluorescens* WCS417r-induced resistance is mediated by an SA-independent signaling pathway, thus concluding that rhizobacteria-mediated ISR and pathogen-induced SAR are regulated by different signaling pathways (Pieterse et al., 1996). Although SAR and ISR work through different signaling pathways, both forms confer resistance in uninoculated systemic tissues, following initial resistance induction at the site of inoculation/colonization; and activation of both forms of resistance can provide plants with even stronger defense (Pruitt et al., 2021).

Along with salicylic acid, two other plant signaling molecules regulate plant defense against microbial attack: jasmonic acid and ethylene. Jasmonic acid/ethylene-dependent signaling pathways are essential for a plant's response to mechanical wounding, pathogen attacks, and herbivore predation (Pieterse et al., 2012). PGPM-mediated ISR is particularly effective against attackers that are sensitive to JA/ET-dependent defenses, including necrotrophic pathogens and insect herbivores (Glazebrook, 2005; Pieterse et al., 2014). The importance of JA and ET in PGPM-mediated ISR has been demonstrated through the using JA and ET signaling mutants in *Arabidopsis*, where induction of ISR by *P. fluorescens* WCS417r was shown to be defective (De Vleesschauwer et al., 2008; Pieterse & Van Loon, 1999). Similar observations have been made in other plant species, such as tomato and rice, as well as with other PGPM, supporting the notion that JA and ET are dominant players in the regulation of the SA-independent systemic immunity conferred by beneficial soilborne microorganisms (De Vleesschauwer et al., 2008; Magotra et al., 2016).

ISR triggered by PGPM is not necessarily associated with enhanced biosynthesis of JA and ET or with massive changes in defense-related gene expression. Instead, ISR-expressing plants are primed for enhanced defense. Priming is characterized by a stronger and faster expression of cellular defense responses that are activated only upon pathogen attack, resulting in an enhanced level of resistance to the infection (Goellner & Conrath, 2008; Liu & Brettell, 2019; Mauch-Mani et al., 2017). The priming phase takes place after priming stimulation, which can be caused by pathogens, beneficial microorganisms, arthropods, chemicals or/and abiotic stimuli, and last until exposure to challenging stress. During this phase, various changes (physiological, transcriptional, and metabolic) occur in the plant after perceiving the stimulus, preparing it for enhanced responsiveness when a challenge occurs (Liu & Brettell, 2019). These changes can manifest as an accumulation of cytosolic calcium and reactive oxygen species (ROS), membrane depolarization, transcriptional modifications, and gene activation, among others (Campos-Soriano et al., 2012; Chaparro-Encinas et al., 2022; Lee et al., 2020). When a primed plant is attacked by a pathogen, enhanced signal transduction and defense responses occur. Primed plants can exhibit potentiated ROS generation (Farooq et al., 2019), callose deposition (Wang et al., 2021), and earlier and/or stronger gene expression in response to a challenge, which is one of the most common responses detected in primed plants (Bakker et al., 2013; Mauch-Mani et al., 2017). Several publications have reported post-challenge increases in gene expression and activities of chitinase, glucanase, phenylalanine ammonia-lyase (PAL), peroxidase, and plant defensin (PDF1.2) genes involved in various steps of the JA signaling pathway (such as lipoxygenase, LOX) JA-responsive gene VSP, as well as in the content of phenolic compounds in plants treated with ISR-eliciting PGPM (Lakshmanan et al., 2013; Liu et al., 2017; Zhang et al., 2015; 2016).

SAR and ISR are not necessarily mutually exclusive, but it has been shown that the activation of SAR can suppress JA signaling in plants, thereby prioritizing SA-dependent resistance to microbial pathogens. Antagonism between the SA and JA response pathways has been observed in *Arabidopsis*, highlighting the potential significance of SA and JA cross-talk in nature (Pieterse et al., 2012). Studies have demonstrated that blocking SA accumulation in *Arabidopsis* during *Pseudomonas syringae* infection leads to the enhanced expression of several JA-induced genes (Spoel et al., 2003). The exact mechanism of this interaction is still not well understood, but it is known that the repression of the JA pathway by SA requires the function of NPR1, although not its nuclear localization (as in SAR). This suggests that NPR1 may mediate SA suppression of JA signaling through a cytosolic function (Pieterse et al., 2012).

The NPR1 protein has also been shown to be required for JA/ET-dependent ISR triggered by PGPM. Pieterse et al. (1998) tested the *Arabidopsis* mutant npr1 and found that npr1 plants were affected in the expression of *P. fluorescens* WCS417r–mediated ISR when infected with *P. syringae* pv tomato, indicating that both types of biologically induced disease resistance are dependent on NPR1. Similar results were found using the *Arabidopsis* mutant npr1-3, which lacks a functional nuclear localization of NPR1 but retains a cytosolic function; and these plants showed reduced amounts of powdery mildew conidia, indicating that the SA-dependent pathway and nuclear localization of NPR1 are not required for mycorrhizal fungus *Piriformospora indica*-mediated induced resistance (Stein et al., 2008). Therefore, the mechanisms of action of NPR1 in the cytosol are not well understood, and further research is needed.

ET also plays a significant role in modulating the outcome of the JA response. When produced in combination with JA, it acts synergistically on the expression of the ERF branch of the JA pathway (regulated by members of the Apetala2/Ethylene Response Factor AP2/ERF family of transcription factors and includes the JA-responsive marker gene Plant Defensin1.2 PDF1.2), whereas it antagonizes the MYC branch (controlled by MYC-type transcription factors and includes the JA-responsive marker gene Vegetative Storage Protein2 VSP2), prioritizing the immune signaling network toward JA/ET-dependent defense signaling associated with resistance to necrotrophs (Dombrecht et al., 2007; Lorenzo et al., 2003; Pieterse et al., 2012). Also, during SA–JA signal interaction, ET plays a critical role, as simultaneous induction of the JA and ET pathway renders the plant insensitive to future SA-mediated suppression of JA-dependent defenses, which may give priority to the JA/ET pathway over the SA pathway during multi-attacker interactions (Leon-Reyes et al., 2010).

The use of PGPM as inducers of systemic resistance in plants is a very attractive tool for producers and growers, as these not only confer better crop yields sustainably but also have the potential to biocontrol and protect the plants, along with many other ecological benefits, which will be detailed in the next sections.

Chapter 8

Omics approaches for detecting action modes of microbial inoculants

Eber D. Villa-Rodríguez[1], Alondra María Díaz-Rodríguez[2] and Sergio de los Santos Villalobos[2]
[1]*Aarhus University, Aarhus, Denmark;* [2]*Instituto Tecnológico de Sonora, Ciudad Obregón, Sonora, Mexico*

The plant microbiome is constituted of a great diversity of microorganisms, including bacteria, fungi, algae, and nematodes, which have profound effects on plant growth and health (Berg et al., 2021). In the last decades, the use of plant-associated microorganisms has gained attention as an alternative to reduce the intensive application of toxic chemicals and fertilizers in agriculture. This is evidenced by the growing rate of the microbial inoculant market in agriculture, with a compound annual growth rate (CAGR) of 15%−18% and a projected value of 10 billion US$ by 2025 (Berg et al., 2021). Although significant progress has been made in understanding the mechanisms by which these microorganisms promote plant growth (including both indirect and direct mechanisms), there are still knowledge gaps that need to be filled to better understand and harness the potential of microbial inoculants (Sessitsch et al., 2019).

In general, bioprospection studies typically begin with the isolation of hundreds of microorganisms from agroecosystems with similar characteristics (such as target crop, plant compartment, and agroclimatic conditions) to the intended application site (Köhl et al., 2011). Subsequently, these microorganisms undergo screening and selection based on well-known plant growth-promoting traits (see Chapter 5) to finally evaluate their performance in promoting plant growth under greenhouse and field conditions (Köhl et al., 2011). While this strategy is useful for identifying potential modes of action underlying plant growth promotion, it is important to note that the in vitro conditions used to assess plant growth-promoting traits may differ from those found in situ, and therefore, the in vitro characterization may not always accurately predict the in situ performance of microorganisms in plants.

New Insights, Trends, and Challenges in the Development and Applications of Microbial Inoculants in Agriculture
https://doi.org/10.1016/B978-0-443-18855-8.00008-4

Despite the demonstrated improvement in crop production through the application of microbial inoculants, it has been observed that their performance can vary across different applications, presumably because the environmental conditions (e.g., biotic and abiotic factors) are not conducible to express their plant growth-promoting effects (Akinrinlola et al., 2018). For this reason, the development of microbial inoculants should be accompanied by comprehensive in situ studies to understand the mechanism of action underlying plant growth promotion. The last could drive the design of novel formulations or agricultural practices aimed to enhance the performance of microbial inoculants.

8.1 Approaches for studying in situ plant growth promotion traits

Plant growth promotion mechanisms exerted by plant-associated microorganisms are generally classified as direct and indirect, according to the type of action on the plant. Phytohormone production, nitrogen fixation, phosphate solubilization, and 1-aminocyclopropane-1-carboxylic acid (ACC) deaminase production are some examples of direct plant growth promotion mechanisms as these have a direct effect on plant nutrition or physiological processes, which trigger plant growth promotion. On the other hand, indirect mechanisms include those that act against plant pathogens, such as the production of antibiotics, lipopeptides, siderophores, and lytic enzymes. The concept of microbiome modulation, a new mechanism of action, has recently been proposed as a consequence of changes in microbial communities induced by the inoculation of plant growth-promoting microorganisms, which result in positive effects such as plant growth promotion and phytopathogen suppression (Berg et al., 2021). The study of plant growth-promoting mechanisms in situ is not an easy task, however, in the past few decades, the development of omics technologies, including genomics, transcriptomics, metabolomics, and proteomics, has played a crucial role in detecting the activity of these traits at different functional levels, such as genes, transcripts, proteins, and metabolites (Fig. 8.1).

8.1.1 Genomic analysis

Since the first microbial genome was sequenced in 1995 using Sanger sequencing technology, a dramatic improvement in DNA sequencing technologies has been made through the development of second/next and third-generation sequencing technologies, which has led to a significant reduction in sequencing costs (Land et al., 2015). The last has enabled the generation of genome sequencing data at a much larger scale, contributing immensely to the understanding of the physiology of PGPM and the identification of genes involved in plant growth promotion activities. Currently, computational resources such as KEGG, SEED, Bio/MetaCyc, GO, InterPro, and RAST allow

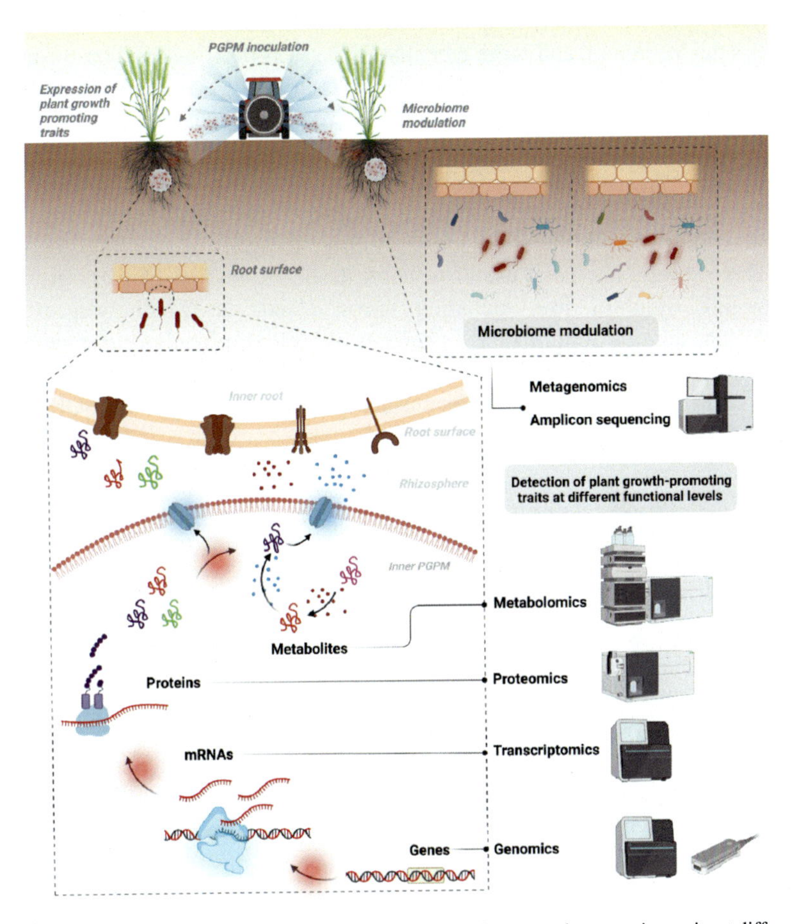

FIGURE 8.1 Application of omics approaches to detect plant growth-promoting traits at different functional levels.

the functional annotation of genomes and the classification of genes involved in different biological processes (Ejigu & Jung, 2020). Nonetheless, these resources have been outperformed by the development of tools such as anti-SMASH (Blin et al., 2021), BiG-SCAPE (Navarro-Muñoz et al., 2020), and PLaBAse (Patz et al., 2021), which allow the mining of genes involved in the biosynthesis of antimicrobial metabolites and plant growth promotion activities from a genome or metagenome-assembled genomes (MAGs). Although nowadays it is possible to know the potential plant growth-promoting traits of microorganisms accurately with their genomic information, their role in plant growth promotion cannot be assumed without experimental evidence.

The development of precise genetic engineering tools and the availability of complete genome sequences has facilitated the knockout of practically any gene of interest, allowing the linkage of resulting phenotypes with gene functions. Over the past few decades, different strategies have been developed for gene knockout, including insertional mutagenesis, markerless deletion using suicide vectors, and more recently, markerless deletions and base editing using CRISPR-Cas systems (Arroyo-Olarte et al., 2021). In the past, genome engineering tools were primarily limited to model organisms; however, with the wide range of available options today, it is possible to knockout genes in nonmodel strains, including PGPM. Gene knockout, followed by phenotypic characterization, is a powerful strategy to explore in vitro and in vivo function of genes (Fig. 8.2).

This strategy has been useful to confirm the expression of plant growth-promoting traits in situ, including nitrogen fixation (Madhaiyan et al., 2013), phosphorus solubilization (Roca et al., 2013), phytohormone production (Idris et al., 2007) and ACC deaminase activity (Jaemsaeng et al., 2018; Peng et al., 2019) (Table 8.1). For instance, *B. amyloliquefaciens* SQR9 is a

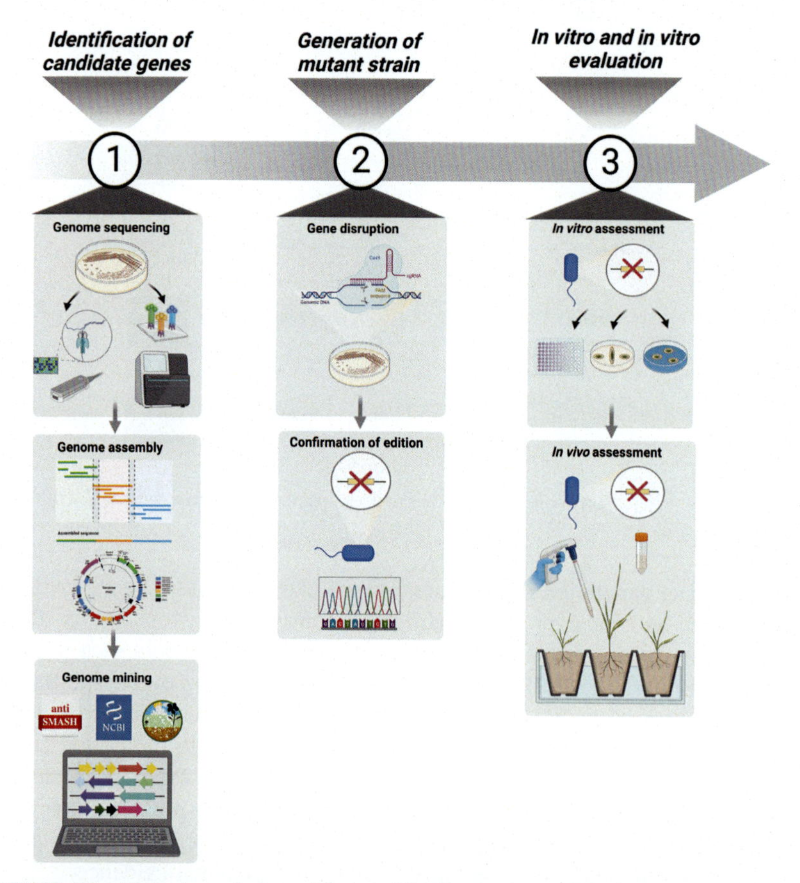

FIGURE 8.2 Genomic analysis workflow to identify plant growth-promoting genes. In this approach, genome mining is conducted to identify potential PGPR genes. Subsequently, mutants lacking those genes are generated using different approaches such as CRISPR-Cas9, markerless deletion, or insertional mutagenesis. Finally, the different mutants are inoculated onto plants to confirm the role of the selected genes in plant growth promotion.

TABLE 8.1 An overview of approaches/strategies in situ expression of plant growth-promoting traits by PGPM.

Approach	Mechanism of action	Strategy	References
Genomic analysis	Phytohormone production	*yhcX* deletion	Yunpeng Liu et al. (2016)
		trpAB deletion	Idris et al. (2007)
	Phosphorus solubilization	*gcd* deletion	Roca et al. (2013)
	Nitrogen fixation	*nifK* deletion	Madhaiyan et al. (2013)
	ACC deaminase production	*acdS* deletion	Peng et al. (2019)
		acdS deletion	Jaemsaeng et al. (2018)
	Antimicrobial production	*srfA* deletion	Chowdhury et al. (2015)
		phzA-H deletion	Chen et al. (2018)
		acbA-J deletion	Xu et al. (2022)
Transcriptomic analysis	Phytohormone production	Untargeted (RNAseq)	Camilios-Neto et al. (2014)
	Nitrogen fixation	Targeted (RT-qPCR)	Yongbin Li et al. (2020)
		Targeted (RT-qPCR)	Wongdee et al. (2018)
		Untargeted (RNAseq)	Camilios-Neto et al. (2014)
	ACC deaminase production	Targeted (RT-qPCR)	Laveilhé et al. (2022)
	Antimicrobial production	Untargeted (RNAseq)	Hu Liu et al. (2021)
		Untargeted (RNAseq)	Kang et al. (2018)
	Other functions (biofilm, motility, chemotaxis)	Untargeted (RNAseq)	Klonowska et al. (2018)
		Untargeted (RNAseq)	Drogue et al. (2014)

Continued

TABLE 8.1 An overview of approaches/strategies in situ expression of plant growth-promoting traits by PGPM.—cont'd

Approach	Mechanism of action	Strategy	References
Proteomic analysis	Nitrogen fixation	Untargeted proteomics	Dicenzo et al. (2019)
	Antimicrobial production	Untargeted proteomics	Qiu et al. (2014)
	Induction of systemic response	Untargeted proteomics	Kierul et al. (2015)
	Other functions (biofilm, motility, chemotaxis)	Untargeted proteomics	Qiu et al. (2014)
		Untargeted proteomics	Kierul et al. (2015)
Metabolomic analysis	Phytohormone production	Targeted metabolomics	Yunpeng Liu et al. (2016)
	Induction of stress tolerance	Targeted metabolomics	Elsakhawy et al. (2019)
	Phosphorus solubilization	Targeted metabolomics	Oteino et al. (2015)
	Induction of systemic response	Targeted metabolomics	Yunlong Li and Chen (2019)
	Antimicrobial production	Untargeted metabolomics	Villa-Rodriguez et al. (2021a)
		Targeted metabolomics	Debois et al. (2014)
		Untargeted metabolomics	Xu et al. (2020)
		Untargeted metabolomics	Millan et al. (2022)

Microbiome analysis	Microbiome modulation—phytopathogen suppression	16S rRNA and ITS amplicon sequencing	Guimarães et al. (2020)
		16S rRNA and ITS amplicon sequencing	Saravanakumar et al. (2017)
		16S rRNA amplicon sequencing	Passera et al. (2020)
	Microbiome modulation—plant growth promotion	16S rRNA and ITS1 amplicon sequencing	Kusstatscher et al. (2020)
		16S rRNA and ITS2 amplicon sequencing	Kaur et al. (2022)
		16S rRNA amplicon sequencing	Wang et al. (2018)

well-known plant growth-promoting bacteria of cucumber. The deletion of the *yhcX* gene in this strain (SQR9Δ*yhcX*), which is an essential gene for auxin biosynthesis, reduced its ability to promote cucumber growth, evidencing that SQR9 promotes cucumber growth by auxin production (Yunpeng Liu et al., 2016). Likewise, *acdS* deletion (Δ*acdS*)—a gene involved in ACC deaminase production—in different plant growth-promoting strains resulted in the loss of plant growth-promotion activity, particularly under salt-stress conditions, demonstrating that these microorganisms induce stress resistance by inhibiting the plant's ethylene production (Jaemsaeng et al., 2018; Peng et al., 2019). On the other hand, this approach has also been useful to elucidate indirect plant growth-promoting mechanisms by disrupting genes involved in secondary metabolite biosynthesis. This can be exemplified by studies where the deletion of gene clusters involved in the synthesis of antimicrobial metabolites, such as phenazines (Chen et al., 2018) anthranilic acid (Matsumoto et al., 2021), herbicolin A (Xu et al., 2022), iturins (Cao et al., 2018), and surfactins (Chowdhury et al., 2015), resulted in the loss of in situ biological control activity against different phytopathogens, such as *R. solani*, *F. graminearum*, *F. oxysporum*, *R. solanacearum*, and *B. plantarii* (Table 8.1).

8.1.2 Transcriptomic analysis

The study of gene expressions (transcriptomics) under certain conditions provides relevant information about the physiological and metabolic state of PGPM, which can be useful to elucidate their mechanisms of action in situ. Early transcriptomic studies were performed with a targeted approach on candidate genes using reverse transcription quantitative real-time PCR (RT-qPCR) (Fig. 8.3). With this approach, gene selections to monitor their expression are commonly made based on prior observations of in vitro plant growth-promoting traits conducted during the screening in the bioprospection stage and genome mining analysis. This strategy has been useful to detect the in situ activity of direct and indirect plant growth-promoting traits by monitoring the expression of genes involved in nitrogen fixation (*nif*) (Yongbin Li et al., 2020; Wongdee et al., 2018), phosphorus solubilization (*gcd, pho, phy*) (Rasul et al., 2021), indoleacetic acid production (*yhcX, yclC, yclB*) (Yunpeng Liu et al., 2016), and antimicrobial metabolites production involved in phytopathogen inhibition (*phlD, phzF, prnD*) (Imperiali et al., 2017). Even though the targeted approach provides evidence of plant growth promotion activities in situ, it lacks the resolution to comprehensibly analyze the transcriptomic landscape and therefore explore novel mechanisms that cannot be discovered following this approach.

The development of microarray DNA hybridization devices facilitated the study of complex transcriptomic responses in living organisms. Despite microarrays making possible the study of the first transcriptomes of plant-associated microorganisms during their interaction with plants, revealing the

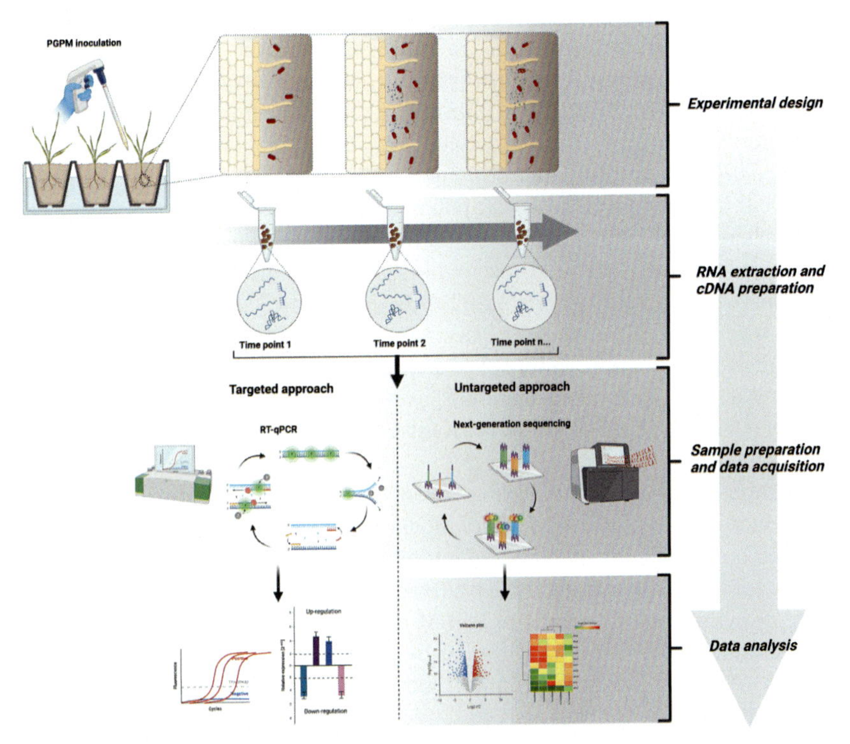

FIGURE 8.3 Transcriptomic analysis workflow to identify plant growth-promoting genes. The experimental design largely depends on the plant growth promotion activity. Bacterial RNA is extracted from the different conditions established (control vs. experimental conditions). For targeted transcriptomics, DNA is depleted from RNA samples and then converted to cDNA. Then, specific primers are used to amplify the pre-selected genes and housekeeping genes to know their relative expression across different conditions. For untargeted transcriptomics, DNA and rRNA are depleted and then sequencing libraries are constructed and sequenced depending on the sequencing platform chosen. Then, reads are mapped against reference genomes. Statistical and gene ontology analyses are performed to identify differentially expressed genes/pathways across different conditions.

complex transcriptomic responses that occur during plant–microbe interactions, the study of PGPM was primarily limited to model strains due to the unavailability of commercial microarray devices (Massart et al., 2015). This limitation was overcome with the introduction of next-generation sequencing (NGS) platforms and RNA sequencing methods (RNA-seq), which surpass microarrays in several aspects such as higher resolution, better detection limits, and lower technical variation (Massart et al., 2015).

RNA-seq has become the standard method to explore transcriptomic responses of PGPM, providing valuable insights into the complex gene regulations that promote plant growth. Similar to RT-qPCR, RNA-seq has proven to be a useful method to identify in situ function of well-known plant growth promotion traits (Fig. 8.3). For instance, using this approach, the overexpression of genes involved in nitrogen fixation was identified in *A. brasilense* during its interaction with wheat roots, demonstrating the functionality of *A. brasilense*'s nitrogen fixation machinery and its role to promote wheat growth (Camilios-Neto et al., 2014). Likewise, analysis of the *Paenibacillus polymyxa* transcriptome revealed the expression of genes involved in the biosynthesis of polymyxin and fusaricidin pepper plants rhizosphere, well-known antimicrobial metabolites produced by this bacterial genus, evidencing that these metabolites were responsible for *Xanthomonas citri* suppression (Hu Liu et al., 2021). Beyond these common plant growth-promoting traits, RNA-seq analysis has revealed genes involved in niche adaptation and competition for space and nutrients, which are crucial activities to interact with plants, and therefore, exert their plant growth-promoting activities (Camilios-Neto et al., 2014; Kang et al., 2018; Klonowska et al., 2018) (Fig. 8.1).

Although an outstanding number of studies have explored the transcriptome of plant–microbe interactions, these studies have predominantly focused on the plant responses to microbial inoculation. In this sense, there is a need for further research on the transcriptional responses and modulation of plant growth-promoting traits of PGPM in situ. Future transcriptomic studies should address the effect of different biotic and abiotic factors on the modulation of plant growth promotion activity and niche competitiveness by PGPM.

8.1.3 Proteomic analysis

The study of the proteome, defined as the complete set of proteins expressed by an organism, provides information about the physiological and metabolic activities of organisms under specific conditions (Fig. 8.4). In this sense, proteomic analysis of PGPM in situ can provide useful information about their mechanisms of action by identifying and quantifying proteins involved in plant growth-promoting traits. Similar to transcriptomics, proteomic approaches can be classified as targeted or untargeted. Although both strategies can use the same instrumental equipment (such as liquid chromatography-mass spectrometry), targeted proteomics emphasizes the analysis of a predefined set of proteins or peptides. On the other hand, untargeted proteomics analyzes the entire proteome without a predefined selection of proteins, offering a more comprehensive overview of the proteome (Massart et al., 2015).

In recent decades, untargeted proteomics studies have been used to determine the impact of plant-associated microorganisms in plants. Through this approach, the influence of these microorganisms in modulating plant metabolic pathways has been demonstrated, providing insights into the plant reprogramming mechanisms induced by PGPM (Khatabi et al., 2019; Perazzolli et al., 2012). However, the number of proteomics studies focusing on PGPM proteomes in plant–microbe interaction studies is considerably lower

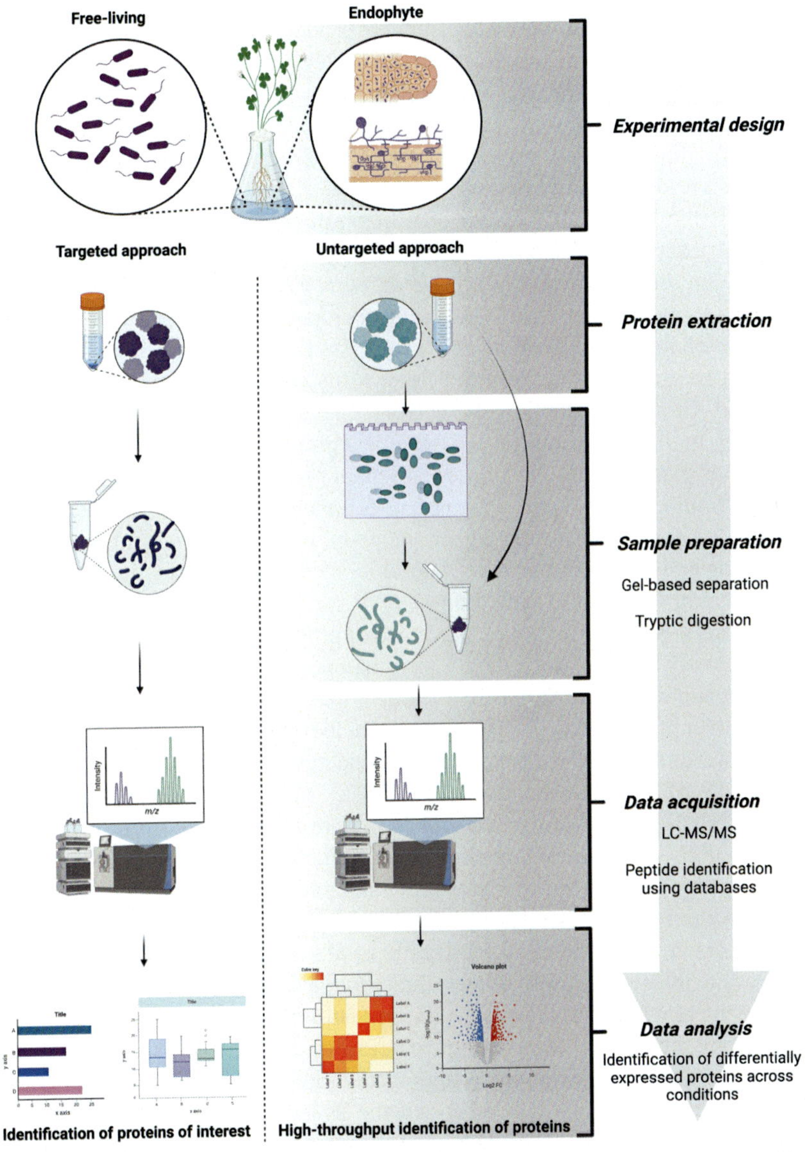

FIGURE 8.4 Proteomic analysis workflow to identify plant growth-promoting proteins. The experimental design largely depends on the plant growth promotion activity. The total protein is extracted from the different conditions established (control vs. experimental conditions). For targeted proteomics, protein samples can or not be digested using proteases. Then, the sample is injected into the mass spectrometry (MS) analyzer, which is focused on detecting and quantifying a preselected set of peptides. Statistical analysis is performed to determine the protein/peptide abundance across conditions. On untargeted proteomics, the digested sample can be either separated on two-dimensional gel electrophoresis (2-DE) or directly analyzed on a liquid chromatography-tandem mass spectrometry (LC-MS^2) analyzer. Peptides are identified by comparing MS spectrums against databases and statistical analyses are performed to identify differentially produced proteins across conditions.

compared to studies on plant proteomes, presumably due to the technical challenges involved in obtaining enough protein extracts from microorganisms, especially in the phyllosphere and rhizosphere compartments. As a consequence, experimental alternatives have been established to mock plant−microbe interactions (e.g., exposing the bacteria to plant root exudates) and study the proteomic responses of PGPM. Even though this experimental setup is not ideal, mainly because it lacks the plant−microbe crosstalk, it has proven helpful in identifying the expression of plant growth-promoting traits and revealing niche competitiveness strategies of PGPM (Table 8.1). For instance, in the biocontrol agent *B. velenzensis* FZB42, an increase in proteins related to plant colonization and niche competitiveness activities (such as motility, chemotaxis, biofilm, and sporulation) was observed when it was exposed to maize root exudates. Noteworthy, the highest fold-change was observed in the production of acetolactate synthase (AlsS), an enzyme involved in the synthesis of the volatile compound acetoin, known as an inducer of plant systemic responses and plant growth promotion (Kierul et al., 2015). Similar results were observed in *B. velenzensis* SQR9 exposed to cucumber root exudates, showing an increase of proteins related to biofilm formation and the synthesis of antimicrobial metabolites (surfactin, bacillaene, and difficidin) (Qiu et al., 2014).

Contrary to free-living microorganisms, proteomic approaches seem suitable to study plant endophytes (e.g., rhizobia and mycorrhiza), since these microorganisms colonize specific plant tissues and their population can be higher than that of free-living PGPM. In the case of rhizobia, untargeted proteomic studies have been helpful to identify proteins involved in the regulation of nitrogen fixation (Dicenzo et al., 2019). Furthermore, this approach has led to the discovery of important proteins involved in rhizobia colonization, symbiotic compatibility, and efficiency (Dicenzo et al., 2019). On the other hand, in mycorrhiza, proteomics has been employed to identify mechanisms that alleviate stress tolerance in plants caused by heavy metals, such as arsenic and cadmium (Chiapello et al., 2015).

Despite the contributions of proteomics to the understanding of mechanisms underlying plant growth promotion, its use is still restricted to a limited group of PGPM. In the past, researchers have prioritized studying plant−PGPM interactions using transcriptomics approaches due to the technical facilities. However, recent studies have reported the asynchrony between mRNA and protein levels, and the influence of external environmental stimuli on the translation process (Sharma et al., 2017), highlighting the importance of proteomic studies in the understanding of biological systems, including PGPM. In this sense, more effort is needed in developing novel proteomic pipelines, prioritizing protein recovery efficiency from plant compartments, as well as bioinformatics/biochemical methods to discriminate between plant and microbial proteins. These advancements will be crucial for studying the influence of biotic and abiotic factors on the regulation of plant growth-promoting traits.

8.1.4 Metabolomic analysis

In the omics pipeline (Fig. 8.1)—integrated by genomics, transcriptomics, proteomics, and metabolomics analysis—metabolomics provides unique avenues for deciphering complex metabolic mechanisms occurring in plant–microbe interactions, as metabolites represent the output of the interrelation among the genome, transcriptome, and proteome in living organisms (Adeniji et al., 2020). A great part of the known plant growth-promoting mechanisms of PGPM is mediated directly by metabolite actions ($<$ 1500 Da). Examples of these include phytohormones (indoleacetic acid), phosphorus solubilization (gluconic acid), and phytopathogen inhibition (lipopeptides). Therefore, it is not surprising that metabolomics represents a good strategy for evaluating in situ plant growth-promoting traits of PGPM. In contrast to other omics approaches, the instrumentation used for metabolomics studies is very diverse, ranging from simple colorimetric analysis using spectrophotometers to complex analysis capable of detecting hundreds or thousands of metabolites using mass spectrometers (MS) or nuclear magnetic resonance spectrometers (NMR). Nonetheless, metabolomic approaches are classified as targeted or untargeted (Fig. 8.5), following the same principle as transcriptomics and proteomics analyses (Adeniji et al., 2020).

Metabolomic analyses have gained popularity in studying the influence of PGPM inoculation on plant metabolome. These studies have been useful to identify the modulation of plants' secondary metabolites and propose potential metabolic pathways influenced by these microorganisms (Etalo et al., 2018). However, at present, the number of reports integrating microbial metabolomes in PGMP–plant interaction is limited compared to plant metabolomes. Despite this, metabolomics has been successfully employed to detect the expression of plant growth promotion traits of PGPM in situ (Table 8.1), including phytohormone production (Yunpeng Liu et al., 2016), phosphorus solubilization (Oteino et al., 2015), and antimicrobial production (Debois et al., 2014). Additionally, untargeted metabolomics has driven the discovery of novel microbial metabolites mediating plant growth promotion. For instance, through untargeted metabolomics, it was identified that anthranilic acid produced by the endophyte *Sphingomonas melonis* confers resistance to the phytopathogen *Burkholderia plantarii*, providing the first evidence of the role of this metabolite in plant growth promotion activity (Matsumoto et al., 2021). In *Bacillus subtilis*, untargeted metabolomics revealed the role of amyloid protein TasA in stimulating melon seed germination, representing the first report of a component of *Bacillus* extracellular matrix in plant growth promotion (Berlanga-Clavero et al., 2022). Taken together, these studies highlight the potential of metabolomics studies conducted during PGPM–plant interactions to identify novel plant growth promotion mechanisms that would otherwise be difficult to discover through in vitro experiments.

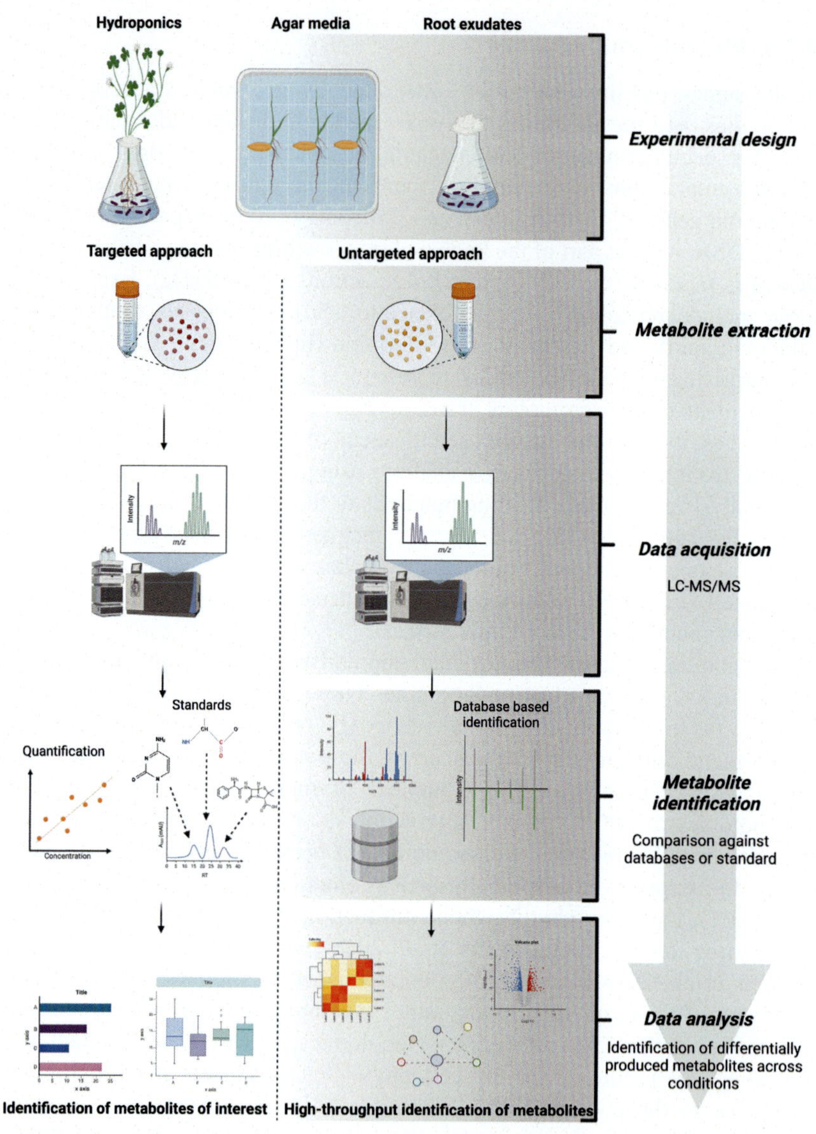

FIGURE 8.5 Metabolomics analysis workflow to identify plant growth-promoting metabolites. The experimental design largely depends on the plant growth promotion activity. Metabolites are extracted from the different conditions established (control vs. experimental conditions) using solvents. For targeted metabolomics, samples are injected into the LC-MS analyzer, which is focused on detecting and quantifying preselected metabolites based on standard curves. Statistical analyses are performed to determine the metabolite abundance across conditions. On untargeted metabolomics, samples are injected in an LC-MS^2 analyzer. Metabolites are identified by comparing MS^2 spectrums against databases and statistical analyses are performed to identify the differential abundance of metabolites across conditions.

One of the major bottlenecks in analyzing microbial metabolomes in PGPM–plant interactions is the capacity to discriminate between plant and microbial metabolites (Adeniji et al., 2020). In contrast to transcriptomics and proteomics, where the discrimination can be achieved through bioinformatic pipelines using genomic data, this is not possible in metabolomics since a great portion of the plant and microbial metabolites are still unknown. Additionally, an alteration in their metabolomes can be expected due to the constant metabolic exchange, increasing the complexity of the metabolomic analyses (Adeniji et al., 2020). As a consequence, most of the metabolomic studies in PGPM have been conducted in vitro so far. Typically, these experiments integrate (1) the identification of metabolites excreted by PGPM growing in axenic cultures and (2) the evaluation of candidate metabolites in plant experiments to identify those mediating plant growth promotion. This strategy has successfully led to the identification of metabolites mediating plant growth promotion via phytopathogen inhibition (Millan et al., 2022; Tsalgatidou et al., 2022; Villa-Rodriguez et al., 2021b; Xu et al., 2020), systemic response induction (Yunlong Li & Chen, 2019), and stress tolerance induction (Elsakhawy et al., 2019). Although the in situ production of these metabolites cannot be assured, this strategy has contributed to the improvement of PGPM bioformulation by optimizing fermentation conditions to improve their production. However, in the future, these metabolomics studies should be complemented with a targeted approach to confirm the in situ production of identified metabolites.

While metabolomics studies of PGPM–plant interaction are still in their early stages, recent advances in metabolomics, such as 3D root cartography platform (Handakumbura et al., 2021), isotopic labeling (Wong et al., 2020), and novel platforms for metabolite dereplication (Aron et al., 2020; Pang et al., 2022), provide promising prospects for the role of metabolomics in unraveling plant–microbe communication and identifying novel metabolites involved in plant growth promotion.

8.1.5 Microbiome analysis

As stated above, the plant microbiome influences multiple plant functions, including germination, growth, and resistance against biotic and abiotic stressors (Berg et al., 2021). These beneficial effects are a result of complex interactions among millions of microorganisms structured by different species, genera, families, and domains (Berg et al., 2021). The concept of a healthy plant microbiome has been extensively discussed in the last few years, although it is not well-defined yet. Generally, microbiomes with high diversity and evenness can be considered healthy, while a decrease in microbiome diversity/evenness is referred to as dysbiosis, which has been linked to increased plant susceptibility to phytopathogen outbreaks (Chen et al., 2020). It is important to note that the plant microbiomes naturally contain phytopathogens as well. In a recent study, the isolation of naturally-occurring phytopathogens from asymptomatic plants, and their re-inoculation in the absence of the microbial community resulted in the appearance of their phytopathogenic lifestyle, denoting the influence of microbiome balance and evenness on plant health (Manzotti et al., 2020).

Pioneering studies on the microbiome employing fingerprinting methods such as DGGE, T-RFLP, and PLFA. While these methods allowed the identification of microbial community changes among samples or conditions, the biological significance of such changes was difficult to interpret due to the low taxonomic resolution (Berg et al., 2021). However, with the development of next-generation sequencing technologies and novel methods for microbiome analysis (e.g., 16S rRNA amplicon sequencing, metagenomics, and metatranscriptomics), it has become possible not only resolve taxonomic composition at the genus/species level but also establish links between taxonomy and biological functions, facilitating the interpretation of microbial communities shifts in the plant microbiome. In recent years, the effect of PGPM inoculation on plant indigenous microbiomes been extensively studied. In general, these effects on the microbiome have resulted in (1) an increase in microbial diversity or evenness (e.g., restoring dysbiosis), or (2) targeted shifts toward well-known plant-beneficial members (Berg et al., 2021). Interestingly, such modulations of the microbiome have been associated with plant-beneficial effects, which has led to the recognition of microbiome modulation as a novel mechanism for promoting plant growth (Berg et al., 2021).

In this sense, analyzing microbial community shifts using 16S rRNA amplicon sequencing or metagenomics is an interesting approach for evaluating the in situ effects of PGPM (Table 8.1, Fig. 8.6). This approach has been used to identify microbiome modulations driven by PGPM inoculation, leading to the suppression of different phytopathogens such as *Fusarium verticillioides* (Guimarães et al., 2020), *Fusarium graminearum* (Saravanakumar et al., 2017), *Rhizoctonia solani* (Passera et al., 2020), and *Phytophthora capsici* (Sang & Kim, 2012). These modulations are presumably linked with the enrichment of bacterial genera capable of producing antimicrobial metabolites, for example, *Peanibacillus*, *Bacillus*, *Pseudomonas*, *Actinomycetes*, and *Burkholderia*. Likewise, microbiome modulation by PGPM introduction has also been related to the improvement of plant growth (Kusstatscher et al., 2020), seed germination (Kaur et al., 2022), stress-resistant (Zhang et al., 2017), and crop yield (Wang et al., 2018). These improvements have been associated with the enrichment of bacterial lineages having members with plant growth-promoting activity, such as actinobacteria, firmicutes, cyanobacteria, and actinobacteria.

Microbiome modulation activity by PGPM has gained attention in the last decade as it provides a pathway for engineering the plant microbiome. However, the mechanisms by which PGPM interact with and modulate

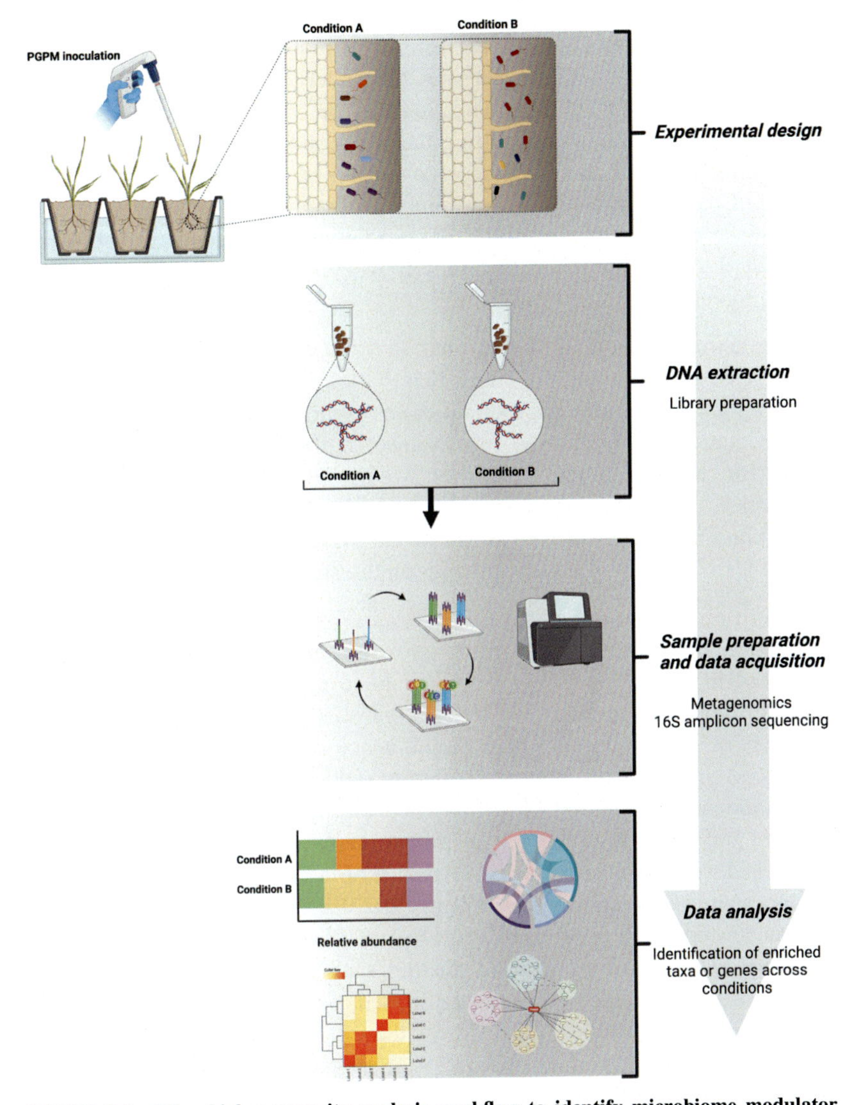

FIGURE 8.6 Microbial community analysis workflow to identify microbiome modulator strains. DNA is extracted from different conditions (with/without PGPM inoculation). Then, sequencing libraries are constructed and sequenced depending on the sequencing platform and method chosen (16S amplicon sequencing or shotgun metagenomics). Gene annotation, taxonomic classification, and statistical analyses are performed to identify the enrichment of genes or taxa across different conditions.

indigenous microbial communities are poorly understood. It is not clear yet whether microbiome shifts resulted from direct microbe–microbe interactions or indirectly by modulating plant root exudation. In either case, further research is necessary to accurately identify the microbial metabolites or genes responsible for the modulation of the microbiome. Currently, it is not feasible to identify microbiome modulator strains in bioprospection stages due to knowledge limitations. Thus, studying the underlying basis of microbiome modulation driven by PGPM inoculation will be crucial to identify potential microbiome modulator strains during the bioprospection stages.

In summary, various omics approaches can be employed to identify the activity of plant growth-promoting traits at different functional levels, that is, gene, transcript, protein, metabolite, or microbiome modulation. The choice of the method to evaluate the in situ activity of PGPM will dependent on many factors, including:

1. The taxonomy of PGPM and their in vitro characterization, which can suggest potential modes of action that PGPM may exhibit in situ. With this information, the most appropriate omics approach can be selected.
2. The purpose of the study, for instance, if the purpose is to monitor a specific trait, a targeted approach would be the most appropriate. In contrast, an untargeted approach would be better suited for exploring multiple traits and the physiology of the PGPM.
3. The team's experience working with omics, which will of course influence the preference for one approach over others.

Despite the expression of plant growth-promoting traits having been confirmed in the past years, it is important to note that a great part of the research has been conducted under controlled conditions (e.g., growth chambers or greenhouses), where the influence of biotic (indigenous microorganisms) and abiotic (climatic conditions, soil type, agricultural practices) factors are not considered. The influence of these factors on shaping the structure and function of indigenous microorganisms is well-known but often overlooked when evaluating the performance of PGPM. The influence of these factors can be exemplified by Kang et al. (2022) and Kusstatscher et al. (2020) who evidenced that the performance of plant growth-promoting traits in the evaluated PGPM is dependent on factors such as nitrogen levels and the indigenous microbiome, respectively. Based on these pioneering studies, future PGPM-plant studies must address the influence of different biotic and abiotic factors on the performance of plant growth-promoting traits. The last should contribute to the design of novel formulations or agricultural practices to improve the performance of PGPM.

Chapter 9

Polyphasic taxonomy of strains in bacterial inoculants

Valeria Valenzuela Ruiz[1], Alejandra Miranda Carrazco[2], Fannie Isela Parra Cota[3] and Sergio de los Santos Villalobos[1]

[1]*Instituto Tecnológico de Sonora, Ciudad Obregón, Sonora, Mexico;* [2]*Departamento de Ciencias Ambientales, UAM Unidad Lerma (UAML), Lerma Estado de México, Mexico;* [3]*Campo Experimental Norman E. Borlaug, INIFAP, Ciudad Obregón, Sonora, Mexico*

An important aspect to address in the formulation of bacterial inoculants is establishing the taxonomic affiliation of the active principles (Fig. 6.1(4)). The use of traditional biochemical techniques, accompanied by a taxonomic allocation through sequencing of phylogenetic markers such as the 16S rRNA for bacteria and archaea domains (Weisburg et al., 1991), is not currently sufficient for a reliable and complete identification of the microorganism in question, because only ~ 1500–1600 bp (base pairs) are evaluated of the ~ 5 million bp in its genome (Land et al., 2015). Thus, a polyphasic taxonomy approach (the combined analysis of genomics, phylogenomic, and biochemical traits) (Das et al., 2014) has been established to accurately affiliate micro-organisms. This approach is necessary to avoid the introduction of potentially pathogenic microorganisms into agroecosystems through microbial inoculants.

A common example of genetic differentiation by the 16S rRNA gene is the *Bacillus (B.) cereus* group, which are gram-positive low-GC-content bacteria, spore-forming, aerobic, facultatively anaerobic, rod-shaped bacteria. This group comprises at least eight closely related species: *B. anthracis, B. cereus, B. thuringiensis, B. mycoides, B. pseudomycoides, B. weihenstephanensis, B. cytotoxicus,* and *B. toyonensis,* which have very similar 16S rRNA gene sequences (Ehling-Schulz et al., 2019). Members of the *B. cereus* group have significant implications on human health, agriculture, and the food industry; for example, *B. anthracis* is the etiological agent of anthrax and an obligate pathogen that poses a threat to human health; *B. cereus* is an opportunistic pathogen known to cause two forms of food poisoning, characterized by either nausea and vomiting or abdominal pain and diarrhea; while *B. thuringiensis* is an insect pathogen widely used in agriculture as a biopesticide due to its

New Insights, Trends, and Challenges in the Development and Applications of Microbial Inoculants in Agriculture
https://doi.org/10.1016/B978-0-443-18855-8.00009-6

production of diverse crystal toxins (Yang Liu et al., 2015). Thus, therein lies the importance and delicacy of high-quality and reliable taxonomic affiliation to avoid mistaking highly genetically relatable bacteria.

In recent years, the introduction of genomic, phylogenomic, and bioinformatics tools has made significant contributions to improving the quality and accuracy of taxonomic affiliations (Paterson et al., 2017). To accomplish a proper affiliation through the polyphasic taxonomy method, full genome sequencing is carried out by next-generation sequencing (NGS) (also known as second-generation sequencing) or third-generation sequencing (TGS) (Fig. 9.1).

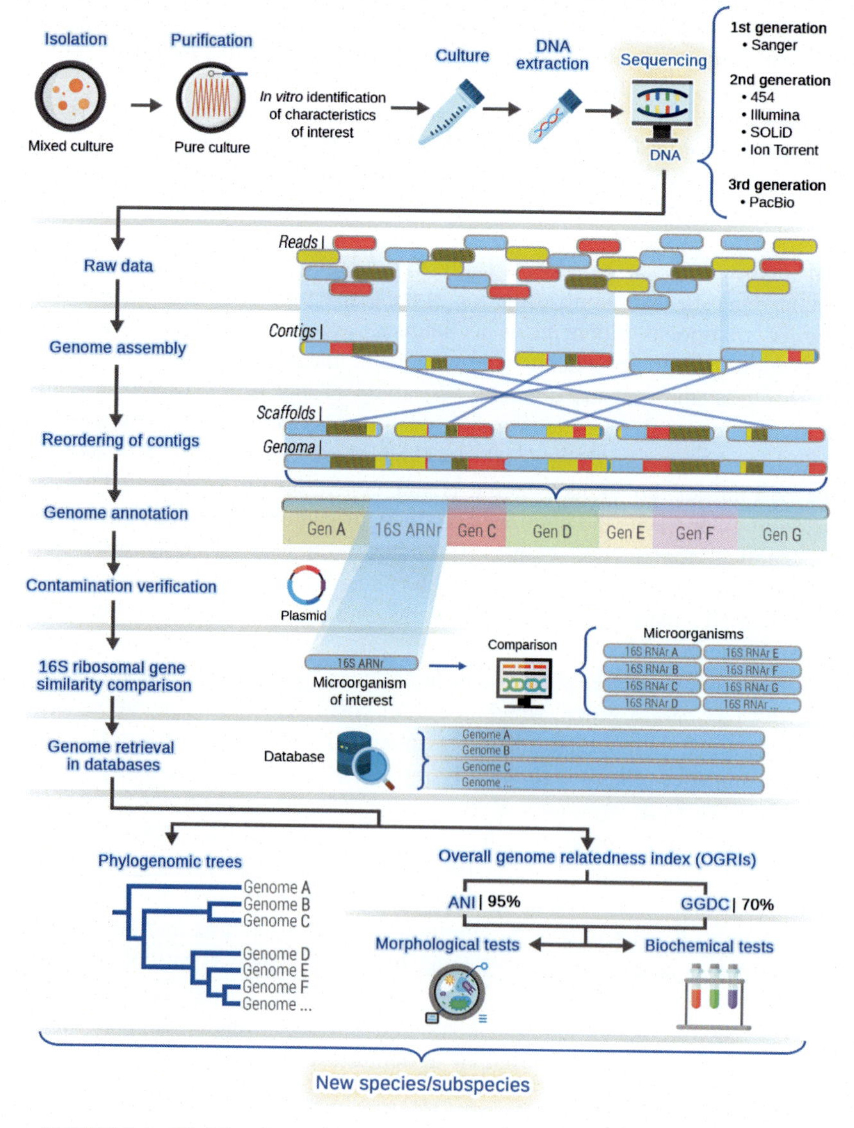

FIGURE 9.1 Workflow for polyphasic taxonomic affiliation of studied bacterial strains.

NGS strategies include various techniques, including sequencing by synthesis (Buermans & den Dunnen, 2014). In this technique, the fragment library serves as the template, by which a new cDNA fragment is synthesized. Sequencing occurs through a cycle of washing and flooding the fragments with known nucleotides in sequential order. As nucleotides are being incorporated into the strand of DNA that is being formed, they are digitally recorded as a sequence (Ambardar et al., 2016; Bahassi & Stambrook, 2014). Current NGS platforms include Roche/454 (Margulies et al., 2005), Illumina/Solexa (Bennett, 2004), Ion torrent, and Sequencing by Oligonucleotide Ligation and Detection (SOLiD), among others, differentiating them from one another.

On the other hand, TGS relies on rapid sample preparation and real-time nucleotide signaling and is generally associated with technologies capable of sequencing single DNA molecules without amplification. The major platforms using TGS technology are Pacific Biosciences (PacBio) single-molecule real-time (SMRT) sequencing, Oxford Nanopore Technologies (ONT) sequencing, and BioNano Genomics (BioNano) sequencing (Staňková et al., 2016). TGS technologies present advantages over NGS platforms, such as; long read lengths, a high percentage of consensus accuracy ($> 99.999\%$ at $30 \times$ coverage depth) which is free of systematic errors, and a low bias of $G + C$ content (Table 9.1) (Wee et al., 2019).

The first step in post-sequencing data processing is a quality control criterion of the raw DNA sequence to exclude or eliminate poor quality DNA reads (Mahamdallie et al., 2018; Xi et al., 2019) through bio-informatic tools like FastQC (Andrews, 2010). Subsequently, after analyzing the quality of the raw reads, sequences not meeting quality criteria and adapter sequences must be discarded; a commonly used tool to do so is Trimmomatic (Bolger et al., 2014). Then, a second quality control revision is done to guarantee the elimination of non-criteria meeting reads.

The next step is to assemble the sequencing reads into larger sequences, known as contigs, this may be carried out through either (1) assembly by comparison or (2) De novo assembly. Assembly by comparison uses the closest phylogenetic species genome to the specie of interest to position and join fragments using the reference genome as a template for read mapping. However, in cases where the phylogeny of the microorganism of interest is unknown, de novo assembly is employed. De novo assembly presents a great challenge due to the presence of repeated sequences, polymorphisms, missing information, or sequencing errors (Robles-Montoya, Valenzuela-Ruiz et al., 2020), so the fragments are joined through bio-informatic algorithms.

Therefore, to successfully assemble the genome of interest, we must also take into consideration the limitations presented in this step due to the sequencing technology chosen. For instance, NGS technologies have small read lengths (< 300 bp), which pose difficulties in de novo assembly. In addition, high/low $G + C$ regions, tandem repeat regions, and interspersed repeat regions are hard to sequence using NGS platforms. Furthermore, de novo genome assemblies may lack entire portions of genomes and missing vital genes, which could be due to fragmentation (Wee et al., 2019). Thus,

TABLE 9.1 Sequencing technologies comparison.

Sequencing technology	First generation	Second generation	Third generation
Method	Fluorescent di-deoxy terminator	Pyrosequencing light emission, fluorescent stepwise sequencing	Fluorescent single-molecule sequencing, proton detection, ligation
Sequencing length	1 kb	35–600 bp	> 1 kb
Cost per kb	\$ 1–2	\$$10^5$–$10^3$	\$$10^4$–$10^3$
Error rate	0.001%–0.01%	0.1%–1%	~ 10%
Error type	Not available	Indel/Mismatch	Indel
Read speed per kb	~ 10 h	~ 10^7 10^4 h	10^7–10^6 h
Sequencing throughput per kb	1 kb	10^8–10^{12} bp	10^9–10^{13} bp
Reads per run	96	100–6 billion	~ 400–600
Advantages	Accuracy	High throughput	Ultralong read length
			Low sequencing-context bias (GC-content or low complexity)
			Uniform coverage along the genome
Disadvantages	Sequencing length is limited to individual genes or short genomes	Short reads limit de novo assembly	Long run time
			Lower output quality
Examples	ABI Sanger	Roche/454	PacBio
		Illumina/Solexa	SMRT
		Ion torrent	ONT
		SOLiD	BioNano

considering that NGS technologies have fewer sequencing errors and TGS technologies facilitate the assembly process, due to longer read fragments, hybrid assemblies using data obtained from both sequencing technologies have recently gained popularity. Hybrid assembly, also known as reference-guided de novo assembly, uses de novo assumptions but also uses related reference genomes to aid in the assembly process (De Souza et al., 2019). The choice of the appropriate assembler strongly depends on the objectives and characteristics of the sequencing project.

There are currently many genomic assembly tools available, among the most commonly cited assemblers include SPAdes (Prjibelski et al., 2020) which works with Illumina or IonTorrent reads and is capable of providing hybrid assemblies using PacBio, Oxford Nanopore, and Sanger reads; SOAPdenovo (Li et al., 2010) which is a novel short-read assembly method that can build a de novo draft assembly for large genomes; another well-reported assembler is Velvet (Zerbino & Birney, 2008), primarily for short-read data assembly. On the other hand, for long-read data, assemblers such as CANU (Koren et al., 2017) and Falcon (Chin et al., 2016) have been reported. CANU supports data from PacBio or Oxford Nanopore, while Falcon is designed specifically for PacBio reads. Besides, Falcon is a diploid-aware assembler, making it particularly suitable for assembling larger genomes. As previously mentioned, hybrid assemblies have gained popularity in recent years; SPAdes, for example, assembles data from second and third-generation sequencing technologies. Other software, like MaSuRCA (Zimin et al., 2013) can perform assemblies between first- and second-generation sequencing platforms, such as 454, Sanger, and Illumina data, through a combination of Brujin graph and overlap-based assembly strategies.

In this way, the assembly must be analyzed in search of errors in either (1) contiguity, which is related to the size and number of contigs and can be equivalent to the number of chromosomes in the organism. These errors may be due to assembler parameters that allow unrelated contigs to be joined or that prevent related contigs from being joined; (2) completeness, which is determined by the content of contigs, and errors in this aspect may originate from sequencing or the assembly process itself; and (3) correctness, which depends on the ordering and location of contigs (Thrash et al., 2020). In this way, to assess these three aspects, there are several quality control servers available. For example, the Computational Geometry Algorithms Library (CGAL) (Rahman & Pachter, 2013), which is a method that is based on the likelihood of an assembly given the original reads. Another example is the benchmarking of universal single-copy orthologs (BUSCO), based on the search for complete, single-copy, fragmented, duplicate, and missing BUSCOs to evaluate assembly quality (Simão et al., 2015). Lastly, the QUality ASsessment Tool (QUAST) (Gurevich et al., 2013), which is mostly based on contiguity measures, such as structural and functional elements, and provides the most information about an assembly when used with an annotated reference genome.

In this manner, to prevent correctness errors the iterative alignment of a genome using a reference genome is highly suggested (Córdova-Albores et al., 2020). This may be done through bio-informatic tools such as Mauve Contig Mover (MCM) (Rissman et al., 2009) which generates a comparative study between draft and reference sequences by ordering draft contigs according to the reference genome. The quality of the reorder is limited by the distance between the sequences, as indicated by the amount of shared gene content between the two organisms. In addition, MCM also orients the contigs in the most likely orientation, and, if annotated sequence features are specified in an input file (i.e., with GenBank format input for thc draft), MCM will output adjusted coordinates ranges for the features.

Furthermore, genome annotation provides us with biological information about our bacteria. This process involves the identification of genes of interest, for example, those related to plant growth promotion, which allows the screening of promising bacteria for agricultural purposes, specifically, as active ingredients in microbial inoculants (Aguilar-Bultet & Falquet, 2015). Genome annotation primarily relies on the detection of homology between newly identified genes/proteins and previously annotated sequences. In this process, newly sequenced genomes are translated and compared against reference databases to identify homologs; and functional annotations are then transferred from those homologs to the query proteins (Lobb et al., 2020).

In the digital age, the transfer of annotations between sequences has become faster than ever before, thanks to a variety of bioinformatic methods and pipelines. Standard approaches include sequence-to-sequence searches, such as BLAST or sequence-to-model searches, such as HMMscan (Ijaq et al., 2015). On the other hand, profile-based methods that use position-specific scoring matrices (PSSMs) or hidden Markov models (HMMs), such as Pfam and the National Center for Biotechnology Information (NCBI) conserved domain database (Finn et al., 2016; Haft et al., 2003; Marchler-Bauer et al., 2011), are among the most sensitive approaches for protein classification. These methods are capable of detecting distant matches to protein and/or protein domain families. Domain families are used to finding matches to building blocks of proteins, such as enzymatic or binding domains, and sometimes allow the transfer of functional information even in the absence of a full protein match (Finn et al., 2016; Lobb et al., 2020).

Diverse bioinformatic algorithms and web servers have been developed in the past years. Among the most used servers are the KEGG (Kyoto Encyclopedia of Genes and Genomes) Automatic Annotation Server (KAAS) (Moriya et al., 2007), and the Rapid Annotation Subsystem Technology (RAST) (Aziz et al., 2008) which uses the Pathosystems Resource Integration Center (PATRIC) (Wattam et al., 2017) to annotate. In addition, command-line annotation pipelines such as MAKER2 (Holt & Yandell, 2011); Prokka (Seemann, 2014), DDBJ Fast Annotation and Submission Tool (DFAST) (Tanizawa et al., 2018) and the Genome Assembly + Annotation Pipeline (GAAP) (Kong et al., 2019) are also commonly used.

In the polyphasic taxonomy of bacteria, genomic cohesion is typically assessed using similarity measures, such as Overall Relatedness Indices (OGRIs), which have been proposed for delineating a species level (Briand et al., 2021). Bacterial species delimitation was originally determined by phenotypic characteristics, such as nutrient requirements, stress tolerance, colony, morphology, and Gram strain, among other analyses. However, due to the ambiguity of these analyses, many strains share certain traits that lead to an erroneous classification. Thus, alternative methods were developed, such as the DNA–DNA hybridization (DDH) method, which was considered the "gold standard" for genotypic circumscription for bacterial species demarcation over the last 50 years (Kim et al., 2014). The DDH technique, described by Schildkraut et al. (1961), involves the duplex formation between denatured DNA from two different organisms to determine their genetic relatedness. The general procedure of DDH consists of fractioning the DNA of the organisms of interest, as well as the DNA of a reference organism, into fragments of 600–800 bp in length. Heat is then applied to dissociate the DNA double strands in the mixture containing the DNA fragments from all the strains. Finally, the temperature is decreased to allow the re-association of such fragments, and the genome similarity percentage is inferred from the melting temperature required, which depends on the level of base pairs matching between strands (Robles-Montoya, Valenzuela-Ruiz, et al., 2020). A value of 70% DDH was proposed by Wayne et al. (1987) as a recommended standard for delineating species. However, DDH procedures are known to be labor-intensive, error-prone, and do not allow the generation of cumulative databases (Hu et al., 2022). Due to these complexities associated with DDH (Rosselló-Móra & Amann, 2015), as well as the increasing availability of whole genome sequence information (Chun & Rainey, 2014; Goris et al., 2007; Richter & Rosselló-Móra, 2009; Sentausa & Fournier, 2013), a range of *in silico* sequence-based measures to replace wet-lab DDH has been proposed. These measures are generally grouped into OGRIs, including the average nucleotide identity (ANI) (Arahal, 2014; Hung & Lee Rutgers, 2016), the genome-to-genome distance (Auch et al., 2010; Chun & Rainey, 2014; Sentausa & Fournier, 2013), maximal unique matches index (Arahal, 2014; Deloger et al., 2009), and tetranucleotide signatures (Chun & Rainey, 2014; Richter & Rosselló-Móra, 2009). Furthermore, dDDH analyses are recommended to be carried out using OGRI values between the type strain of the proposed species and type strains of related species that show $>$ 98.7% 16S rRNA sequence similarity.

One of the most extensive genetic markers still used for phylogenetic purposes is the 16S rRNA gene, which for many years was the base of bacterial classification along with morphological and biochemical characteristics. The 16S rRNA gene has been widely applied in the taxonomy field because it's a highly conserved region present in all bacteria with at least one copy. This allows a simple identification through PCR and provides taxonomic insights related to the family, genus, or in few cases, species assignation of bacterial strains (Mizrahi-Man et al., 2013).

Online bioinformatic servers, such as the EzBioCloud database and 16s-based ID (https://www.ezbiocloud.net/identify), offer similarity-based searches against quality-controlled databases of 16S rRNA sequences against all prokaryotic names in the database. These servers compare the draft or complete 16S rRNA sequence from an isolated strain against the type strain of all prokaryotic species, using an $> 98.7\%$ similarity threshold value (equivalent to the 70% identity of DDH) (Kim & Chun, 2014). The genome of each matching strain above the threshold value is then downloaded for downstream analysis and comparisons that validate the genomic affiliation. However, due to the genetic variation between species of some bacterial genera, these strategies (phenotypic characteristics, DDH, 16S rRNA gene) often lead to low or poor phylogenetic and similarity resolution (Yang Liu et al., 2017). As a result, *in silico* tools have been developed using bioinformatic methods, where these genomic sequences are used to predict the phenotype of the sequenced strain, as well as different phylogenetic and molecular approaches. Among these, several overall genome-relatedness indices (OGRI) have emerged to replace the DDH standard (Richter et al., 2016).

Among various ORGIs, the average nucleotide identity (ANI) has been the most widely used. ANI, as defined by Goris et al. (2007), is a pairwise measure of overall similarity between two genome sequences. ANI values can be obtained using either the Basic Local Alignment Search Tool for nucleotides BLASTn (ANIb) or the MUMmer (ANIm) software (Richter & Rosselló-Móra, 2009), with the former being widely used for taxonomic purposes (Chun & Rainey, 2014). These algorithms, whether using BLAST or MUMmer, calculate ANI values with directional specificity, meaning that when a pair of genomes are compared, the calculated ANI value can be different depending on which genome was selected as the query and which is the subject, although the differences are minor for most cases (Yoon et al., 2017). ANI is calculated from the two genome sequences (of the query and subject strains) as follows: first, the genome sequence of the query strain is divided into 1020 bp-long sequences (fragments). Second, each fragment is searched against the whole genome sequence of the subject strain using NCBI's BLASTn program (Altschul et al., 1990). In this process, the BLASTn calculates nucleotide identity between fragments of the query strain and the genome of the subject strain. The average nucleotide identity is the mean of these nucleotide identity values between the genomes; where only those with at least 30% similarity are considered. The cutoff value typically used to identify bacterial prokaryotic species is $\geq 95\%-96\%$ (Moriuchi et al., 2019; Varghese et al., 2015), while the subspecies cutoff value has been reported to be $\geq 98\%$ (Meier-Kolthoff et al., 2014; Richter et al., 2016). In simpler terms, the ANI value represents the mean identity percentage calculated (Chun & Rainey, 2014). In 2016, an improved ANI algorithm called Average Nucleotide Identity by Orthology (orthoANI) was proposed (Lee et al., 2016). This new algorithm was introduced because DDH values between two strains are often not the same (Johnson & Whirman, 2007; Tindall et al., 2010). This is the average of identity values among all orthologous fragment pairs between two genomes (Robles-Montoya, Valenzuela-Ruiz, et al., 2020).

An alternative to ANI is the digital DDH (dDDH), Genome-to-Genome Distance Calculator (GGDC), which has also been widely used for taxonomic purposes (Chun et al., 2018). GGDC is an OGRI based on distance relationship, where the genome sequences are aligned with each other, obtaining a set of high-scoring segment pairs (HSPs) or maximally unique matches (MUMs), titled "intergenomic matches," from which specific distance formula is calculated to transform HSPs data into the GGDC value, with an established cutoff value of $\geq$ 70% (Meier-Kolthoff et al., 2013). The general working method is divided into three main steps: the determination of a set of HSPs or MUMs between two genomes, the calculation of distances from these sets, and the conversion of these distances into percent-wise similarities analogous to DDH. Where, similarities between query and reference genomes are determined by using tools for nucleotide-based sequence similarity searches, such as NCBI-BLAST (Altschul et al., 1990), WU-BLAST (Altschul et al., 1990), BLAT (Kent, 2002), BLASTZ (Schwartz et al., 2003), and MUMmer (Kurtz et al., 2004). GGDC presents three algorithms for comparing two genomes (a query genome and the reference genome to compare it to all of those whose 16S rRNA were $\geq$ 98.7%), in formula 1, the length of the HSP is divided by the length of the genome; differing form formula 2, where the sum of all identities found in the HSPs is multiplied by 2 and divided by the sum of the total length of the HSPs of both genomes; and formula 3, where the sum of all the HSP identities is divided by the length of the entire genome. However, it is recommended to rely on formula 2 since it is independent of the length of the genome and is suitable for draft genomes. Furthermore, the distance functions of GGDC can also cope with heavily reduced genomes and repetitive sequence regions. Some of them are also very robust against missing fractions of genomic information (due to incomplete genome sequencing). GGDC operates on the same scale as wet-lab DDH values, which makes comparisons much easier.

To confirm the taxonomic affiliation of specific strains, whole genome-based phylogenetic trees are recommended, which represent the divergence in evolutionary relationships among species. Bertels et al. (2014) designed the Reference sequence Alignment-based Phylogeny (REALPHY), which is based on sequence mapping (reference and query genomes) to construct multiple sequence alignments from which phylogenetic trees are inferred. Furthermore, morphological, biochemical, and metabolic traits are analyzed to support the taxonomic affiliation at the species or subspecies level. Thus, the polyphasic taxonomy approach serves as a tool to guarantee the safety of microbial inoculants concerning to the ecological, human, and environmental health of agroecosystems.

Recently, Meier-Kolthoff & Göker (2019), published the Type (Strain) Genome server (TYGS), a web server for genome-based prokaryote taxonomy connected to a large and continuously growing database of genomic, taxonomic, and nomenclatural information. TYGS infers genome-scale phylogenies for species and subspecies boundaries from user-defined and automatically determined closest-type genome sequences. It also provides an integrated approach to genome-based taxonomy by joining features such as a comprehensive database of type strains genomes of species and subspecies with validly published names, automated detection of closest neighbors of query genomes, comprehensive access to nomenclature, and whole-genome-based methods for phylogeny and classification. In general, the TYGS compares query genomes against its database of type (strain) genomes, then infers phylogenetic trees, performs classification at the species and subspecies level, and reports differences in genomic G + C content. These results are obtained through the extraction of the 16S rRNA gene sequences from user-defined genomes, followed by a pairwise nuclear ribosomal small subunit rRNA gene (SSU) BLAST against type strains, the distance calculation, and the identification of the closest type strain, as well as, the high-throughput calculation of missing genome comparisons through the Genome Blast Distance Phylogeny (GBDP), the prediction of dDDH values and their confidence intervals, displaying all this data in a phylogenomic tree.

It is highly recommended to support the previously mentioned genomic analysis with morphological and biochemical characterization, to properly taxonomically affiliate a bacterial species or subspecies. Vandamme et al. (1996) describe phenotypic methods as those that are not directed toward DNA or RNA; including morphological, physiological, and biochemical properties of the organism. These methods involve the application of analytical techniques to gather information on various chemical constituents of the cell to bacterial classification. Traditional phenotypic tests include assessing growth requirements, the growth stress response regarding salinity levels, pH, and temperature, as well as, the susceptibility toward different kinds of antimicrobial agents, among others (Prakash et al., 2007). However, an important aspect to consider is the conditional nature of gene expression wherein the same organism might show different phenotypic characteristics in different environmental conditions. Therefore, phenotypic data must be compared with a similar set of data from the type strain of closely related organisms. Generally, morphological, physiological, and biochemical features alone provide very limited information on genetic relatedness, however, these alongside genomic analysis provide strong evidence of a species taxonomy affiliation.

Furthermore, the morphology of a bacterium includes both cellular characteristics (shape, endospore, flagella, inclusion bodies, Gram staining) and colonial features (color, dimensions, and form) (Vandamme et al., 1996). Other chemotaxonomic markers are commonly evaluated for taxonomical purposes; such as cell wall composition, cellular fatty acid content, isoprenoid quinones, whole-cell protein analysis, and polyamines, among others.

On the other hand, biochemical characterization is widely used for bacterial identification. Some tests are routinely used for many groups of bacteria, including oxidase test, nitrate reduction, amino acid degrading enzymes, fermentation or utilization of carbohydrates, siderophore production, indole acetic acid production, among others (Morales et al., 2021). Most laboratories today use either commercially available miniaturized biochemical test systems or automated instruments for biochemical tests and susceptibility testing. The kits usually contain 10–20 tests. The test results are converted to numerical biochemical profiles, which are then identified using a codebook or a computer. Carbon source utilization systems with up to 95 tests are also available. Most identification takes 4–24 h. Biochemical and enzymatic test systems for which databases have not been developed are used by some reference laboratories.

The integration of genomic analysis, and morphological and biochemical characterization, resulting from a polyphasic taxonomic approach, provides the necessarily supported information to taxonomically affiliate bacterial species. This integration allows us to have a better understanding of the microorganisms in general, including evidence about their biosafety, close analysis of agroecological dynamics, and the identification of novel species with great potential for agricultural bioprospecting. Having a genomic background, obtained through gene identification, will not only unravel potential genes of agronomical interest, such as those related to plant growth promotion, and support the identification of PGPB for their further inclusion as active ingredients in microbial inoculants, but also provide additional biological information that can be used to generate specific formulations to improve the growth of the bacteria in culture medium according to their growth requirements. This includes the addition of specific compounds to enhance the production of metabolites of agricultural interest.

Chapter 10

Microbial inoculants production systems

María Fernanda Villarreal-Delgado[1], Alondra María Díaz-Rodríguez[2] and Sergio de los Santos Villalobos[2]
[1]*Sartorius de México, Azcapotzalco, Ciudad de México, Mexico;* [2]*Instituto Tecnológico de Sonora, Ciudad Obregón, Sonora, Mexico*

The development of microbial inoculants begins with the isolation, selection, characterization, and validation tests of effective microorganism and concludes with critical technological steps that are essential for the success of the microbial inoculant: the mass production process and formulation procedure (Vassilev et al., 2015).

Microbial inoculants are produced through the fermentation process, which involves the cultivation of microorganisms in a specific nutrient medium to produce specific metabolites or biomass. The development of a synthetic and optimum medium is crucial, particularly for submerged cultivation of axenic organisms, to achieve high-density concentrations and eliminate any traces of unutilized complex sources, thus preventing contamination of the microbial inoculant throughout the supply chain until delivery (Vassilev et al., 2015).

The optimum media and culture conditions are ideally studied for each potential strain to maximize the production process, which depends on the specific fermentation objective. These objectives can include the (1) production of microbial biomass, (2) bioconversion of microbial substrate, (3) production of primary or secondary metabolites, or the (4) production of enzymes (Suthar et al., 2017). Analyzing one variable at a time in a fermentation process can be laborious, time-consuming, and often lacks predictive value. To overcome these limitations, statistical models can be used to optimize the fermentation conditions, minimizing the number of experiments and obtaining the ideal process conditions taking into account the interaction between the factors (Pérez Peñaranda et al., 2019). For example, Posada-Uribe et al. (2015) achieved 8.78×10^9 CFU/mL and sporulation efficiency of 94.2% in *Bacillus subtilis* strains by using Placket Burman screening, and further optimization using a full factorial and central composite design with the statistical software Design Expert 8.0.7.1. (Stat-Ease, Inc., Minneapolis, USA). Thi Nguyen and

New Insights, Trends, and Challenges in the Development and Applications of Microbial Inoculants in Agriculture
https://doi.org/10.1016/B978-0-443-18855-8.00010-2

Tran (2018) achieved to increase enzyme concentrations (such as glucose isomerase at 0.274 U/mg biomass) in *Bacillus megaterium* cultures by employing the Plackett–Burman design and response surface methodology in MODDE 5.0 software (Umetrics, Umeå, Sweden).

The main nutrients that usually have a significant impact on the optimization of the fermentation process are those required for the development of microorganism: carbon, from which is derived the energy efficiency of microbial growth, while nitrogen is another essential nutrient involved in synthesis functions, including the genetic material and synthesis of amino acids; likewise, phosphorus and potassium have been related to activation processes in catabolism and anabolism (Dauner et al., 2001). Furthermore, using a culture medium with chemically defined ingredients is of great importance to obtain reproducible and homogeneous cultures (Monteiro et al., 2014). This ensures consistency in the composition of the medium, allowing for reliable and standardized results in the fermentation process.

Another factor that impacts the use of nutrients is their availability in the culture media. The most commonly used method for liquid media sterilization is heat sterilization. However, depending on the process temperature and contact time, the application of heat frequently leads to changes in certain starting materials, which may subsequently have negative effects on the fermentation process. These effects can include (1) the caramelization of sugar solutions; (2) the denaturing of proteins, which are typically used as sources of nitrogen; (3) the inactivation of various vitamins and other essential growth substances; (4) the reaction of aldose sugars with amino acids and other compounds containing amino groups; (5) the polymerization processes involving unsaturated aldehydes; and (6) the hydrolytic cleavages (Berovič, 2011). These changes can alter the composition and properties of the culture media, potentially impacting the growth and metabolic activities of microorganisms during fermentation. Therefore, careful consideration should be given to the sterilization process to minimize any adverse effects on the fermentation process.

One of the strategies validated by the PDA (Parenteral Drug Association) in the original Technical Report No. 26 (1998), is the use of sterilizing filtration. This method allows for the physical removal of microorganisms from chemically and thermally sensitive liquids, thereby avoiding the denaturation of nutrients and preserving their viability in the medium. It involves the use of 0.22 μm membrane filters (Kviat et al., 2008).

Another factor crucial for the successful application of fermentation technology that needs to be considered in the optimization process is the operational conditions, which also impact microbial metabolism. Adequate aeration, which supplies oxygen requirements, is vital for most metabolic processes of aerobic organisms, as oxygen serves as an electron receptor. On the other hand, maintaining a constant pH value during fermentation, typically neutral, benefits the development of biomass, particularly in species such as *Rhizobium, Azotobacter, Azospirillum, and Bacillus* (Suthar et al., 2017). In addition, temperature is one of the most important factors in metabolite biosynthesis and can either accelerate or decrease development times in the culture (Du et al., 2012).

After the selection of a suitable strain, medium, and optimal growth conditions at the laboratory level, the next step is scale-up, which is typically carried out in two stages: pilot-scale production and large-scale production; this involves using bioreactors of different sizes (Bissonette et al., 1986).

The bioreactor is defined as a vessel designed for conducting biological reactions and is used to culture aerobic cells, enabling the production of high cellular mass and quality compounds (Bhatia & Bera, 2015) (Fig. 10.1).

Stirred-tank bioreactors have been widely used due to their numerous benefits, including efficient fluid mixing and oxygen transfer, precise temperature control, low operating costs, compliance with current good manufacturing practice requirements, easy scalability, and the availability of alternative impellers. The previous limitations of this technology, such as high energy requirements, high shear, concerns about sealing, and size limitations due to motor size and shaft length and weight, have been overcome through an understanding of the specific requirements of different types of cultures and target markets (Vikrant et al., 2018). Over the past two decades, advancements in bioreactor designs, the development of automatic control systems for mainlining parameters such as pH, aeration, agitation, and foaming, and mathematical modeling have made fermentation technology more precise, user-friendly, and conducive to microbial fermentation processes (Sartorius, 2022).

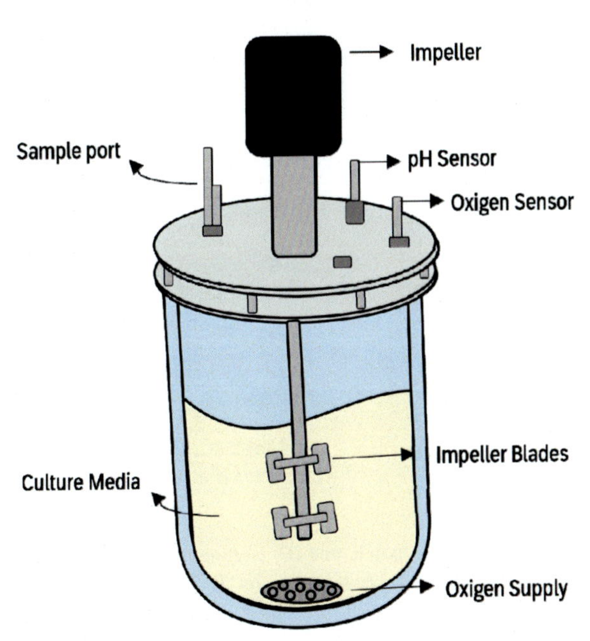

FIGURE 10.1 Schematic representation of a stirred tank bioreactor. Showing the main components of microbial cultivation.

The stirred-tank bioreactors have been successful in bacterial strain production. Monteiro et al. (2014) achieved significant results using Biostat B 2L bioreactors with *Bacillus* strains, reaching 6.3×10^9 spores/mL in batch culture, and 3.6×10^{10} spores/mL in fed-batch culture. These quantities are considered high compared to the optimized culture ranges reported in the literature, which are typically between $7.0–8.2 \times 10^9$ spores/mL (Ooijkaas et al., 1999; Posada-Uribe et al., 2015). In addition, Ramlucken et al. (2021) reported concentrations of 10^{10} CFU/mL with sporulation efficiencies of 80% in their six *Bacillus* strains using batch cultures in Biostat C 30L stirred-tank bioreactors. They collected data points for all process parameters on-line using MFCS software (Sartorius BBI systems, Germany).

Based on the process optimization and systems technology, the fermentation process can be classified into three systems: batch, fed-batch, and continuous process (Fig. 10.2).

- Batch fermentation is a discontinuous process in which nutrients are supplied to the microorganisms only once at the beginning of the fermentation. Once the nutrients are depleted and the product of interest (i.e., biomass,

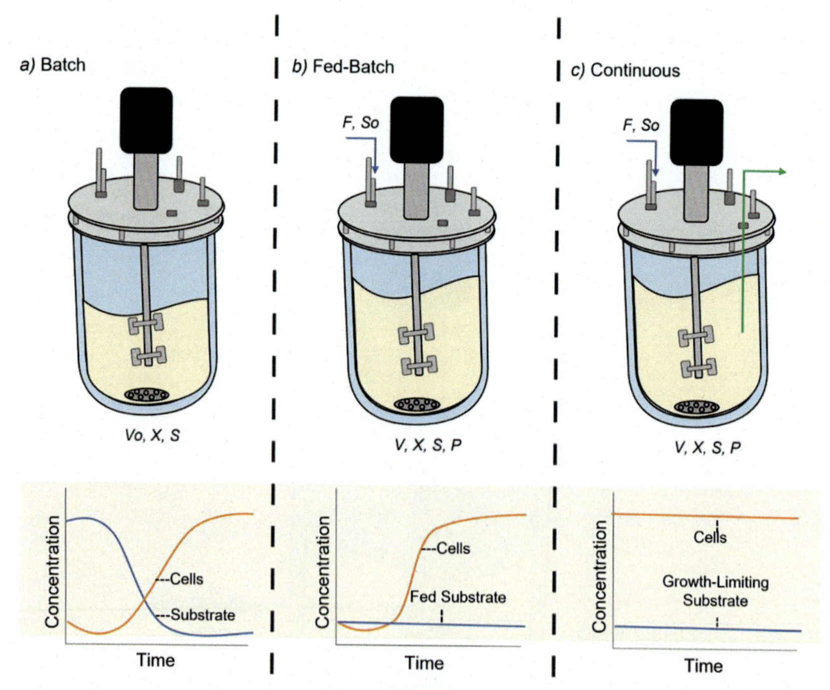

FIGURE 10.2 (A) batch, (B) fed-batch, and (C) continuous cultivation schematic. *F*, Volumetric feed rate at which nutrients are added (L/h); *P*, Concentration of product (g/L); *S*, Instantaneous substrate concentration in the bioreactor (g/L); *So*, Substrate concentration in the feed; *V*, Final volume of the medium in the bioreactor (L); *Vo*, initial volume of the medium in the bioreactor (L); *X*, concentration of biomass (g/L). (Srivastava and Gupta, 2011).

primary or secondary metabolites, etc.) is produced, the entire content of the fermenter tank is harvested for the next step of the processing (Suthar et al., 2017).

- In fed-batch fermentation processes, the rate of nutrient addition of a limiting nutrient is controlled to regulate the reaction rate, and the product is typically harvested at the end of the batch. This fermentation strategy extends the exponential growth phase and maximizes the utilization of the limiting nutrient for the production of the desired product (Srivastava & Gupta, 2011).
- Continuous fermentation involves a continuous supply of nutrients to the microorganisms at a fixed rate, while the products are continuously removed from the fermenter tank. This maintains the microorganisms in the exponential growth phase, resulting in higher production compared to batch fermentation (Bakri et al., 2012).

To ensure the commercial relevance of a bioproduct, the production process must be scalable in mass, maintain high productivity, and achieve high product recovery during harvesting. Likewise, as mentioned, the basis for obtaining high cell densities depends on the optimization process (culture medium and operating conditions), technology, and fermentation strategy (bioreactor and type of culture; batch, batch-fed, or continuous), as well as the inherent capabilities of the microorganism in question for mass production. To achieve mass production, it is necessary to scale up the process from laboratory to plant pilot or industrial-scale bioreactors. During scale-up, it is important to consider certain criteria to ensure that the yield of the product remains comparable at different scales. These criteria include mass-transfer coefficient (kLa), power per unit of liquid, gas hold-up, Reynolds number, and Peclet number, among others (Anaya & Pedroza, 2008).

In the fermentation of aerobics microorganisms, oxygen plays an important role in biomass and metabolite production (Micheletti & Lye, 2006). Usually, kLa is the preferred criteria tool for a convenient scale-up. Flores et al. (1997) found a relationship between spore productivity, which is dependent on the oxygen concentration, and kLa, maintaining this parameter at the bioreactor pressures of 14 and 1100 L which, in addition to presented geometric differences, in both scales were able to achieve concentrations of at least 4×10^9 spores/mL of *Bacillus thuringiensis* in batch culture.

High concentrations of the product of interest (*i.e.*, biomass, metabolites, among others) and high yields are important as they impact the production cost, capital utilization efficiency, and account for any losses in subsequent downstream processes such as cell harvesting and bioproduct formulation.

The final fermented culture comprises a variety of molecules, including biomass, spores, enzymes, metabolites, and other residual solids. Therefore, the molecule of interest must be efficiently recovered for subsequent formulation steps. The recovery process varies significantly depending on the molecule and the production scale of the final product. Key factors to consider for the downstream strategy are the process yield, physical characteristics of the product, and the desired final concentration of the product.

Currently, methods such as ultracentrifugation, tangential flow filtration, and traditional filtration have proven to be efficient for the recovery of various biotechnological products of interest (Brar et al., 2006). For example, tangential flow filtration (TFF) is a rapid and efficient method for the separation and purification of biomolecules (Sartorius, 2022). This technique relies on the size of the biomolecule of interest, the pore size of the filter, the transmembrane pressure of the system, and the creation of a turbulent flow where the sample is recirculated through the system; those molecules that are smaller than the pore size pass through the filter, while larger molecules remain in the concentrate. Orozco-Alvarez et al. (2013) reported a 10× concentration of *Beauveria bassiana* spores using TFF with 0.1 μm or 100 kDa cartridges. Namvar et al. (2013) reported a >80% recovery of *Bacillus* endospores in different matrices through TFF with a pore size of 0.2 μm.

Once the microorganisms or biomolecules of interest have been harvested, the microbial inoculant can be formulated (e.g., powder/suspension, aqueous or oil concentrate, spraying powder, or granulates). It is recommended to incorporate a postharvest stabilizer adjuvant before formulation to prevent spore mortality or germination (Brar et al., 2006).

Chapter 11

Bioformulation of bacterial inoculants

Jonathan Rojas-Padilla, Alondra María Díaz-Rodríguez and Sergio de los Santos Villalobos
Instituto Tecnológico de Sonora, Ciudad Obregón, Sonora, Mexico

Numerous rhizosphere bacteria and fungi, including strains of *Bacillus, Azospirillum, Pseudomonas, Rhizobium, and Streptomyces*, have the potential to directly increase agricultural productivity, shield plants from diseases, and help plants against abiotic stress. The commercial potential of such strains can primarily be determined by the bioprospection of plant growth-promoting microorganisms (PGPM) or bacteria (PGPB), evaluation of their modes of action, and laboratory and/or small-scale efficacy testing. For the successful commercialization of beneficial microorganisms for use in agriculture, the development of the production process, efficacy testing of the product, and its subsequent registration are crucial milestones.

After the biomass production process of the microorganisms of interest for the development of the microbial inoculant, the product should undergo a formulation process. A formulation is usually a combination of active substances from a formed product and inert (inactive) components (http://npic.orst.edu/fastsheets/formulations.html). However, as far as bioformulation is concerned, there is no uniform definition, and several authors define it in their way. Chaudhary et al. (2020) define bioformulation as the preparation of selected beneficial microorganisms with a suitable carrier that can provide stabilization and protection of strain during transport and storage. As described in Chapter 6, the development of microbial inoculants involves many steps. Firstly, is the isolation of the microorganism from soil, rhizosphere, or plant. It is recommended to seek native microorganisms for better adaptability in the bioformulation application. Once the pure strains are isolated, the next steps include identification, metabolic characterization, and conducting in vitro and in vivo efficiency tests. The efficiency tests should be scaled correctly, starting from germination and progressing to growth chamber and greenhouse experiments. This allows the selection of the best strains for their formulation as

New Insights, Trends, and Challenges in the Development and Applications of Microbial Inoculants in Agriculture
https://doi.org/10.1016/B978-0-443-18855-8.00011-4

active ingredients, evaluation of the formulated product, and mass production. The last steps are the conduction of field trials under real conditions, the patent of the product with the final formulation, and its correct registration according to the legislation of each country (Fig. 11.1).

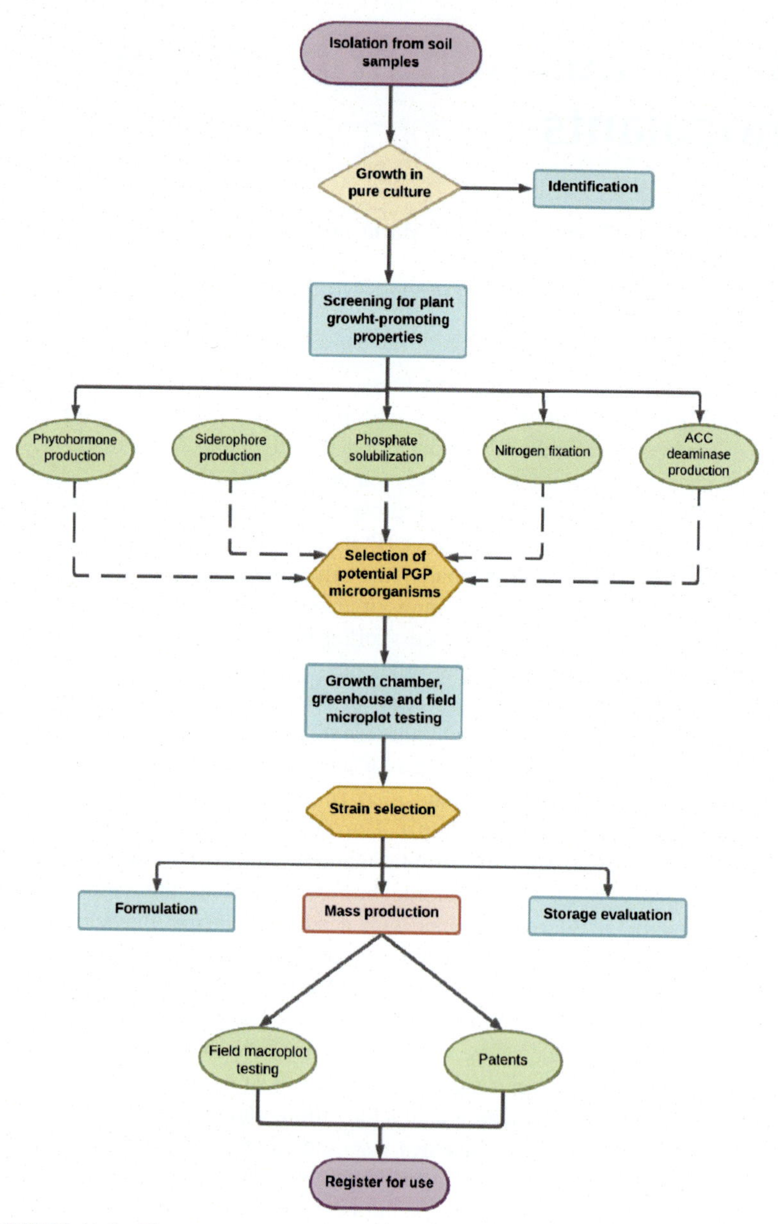

FIGURE 11.1 The procedure proposed for the development of efficient bioformulations.

After the biomass production process of the selected PGPM/PGPB for the development of microbial inoculants, the product should undergo a formulation process in order to ensure the application of the required number of viable microbial cells. The large-scale production of bacteria using bioreactors is a quite common practice in developed countries (see Chapter 10) (Bhatia & Bera, 2015). The formulation process involves the choice of a suitable carrier and the growth of the culture on a large scale with a profitable cost/benefit ratio. Bacterial inoculants can be formulated as liquid or solid-based carriers and can contain a pure or mixed culture (Fig. 11.2) (Alori & Babalola, 2018). During the production, the strain should be able to grow in an artificial medium, survive on carriers, survive when applied to seeds, plants, and soil, and be compatible with agrochemicals products that might be applied to the crops (Herridge, 2008; Herrmann & Lesueur, 2013). For bacterial inoculants, selected bacterial strains are typically cultivated in liquid broth to reach high population levels. The composition of the media and the culture conditions (temperature, pH, moisture content, agitation, and aeration) are directly related to the specific strain's characteristics (Herrmann & Lesueur, 2013; Malusá et al., 2012).

The microbial strain and formulation of inoculants are two aspects that affect the effectiveness of inoculation technology. In practical terms, the success of an inoculant depends on its formulation. In addition, to develop a suitable inoculant, the interests of farmers and manufacturers must be taken into account. Farmers expect the product to be easy to use in the field without requiring changes to their usual practices, and to be compatible with disinfectants and pesticides used on seeds and plants (Bashan et al., 2014).

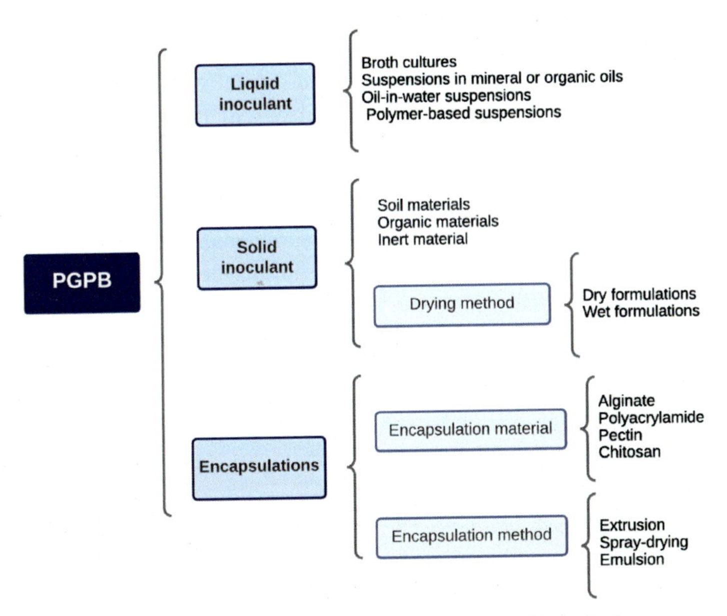

FIGURE 11.2 Types of bioformulations for sustainable agriculture.

In practice, the carrier serves as the delivery vehicle for live microorganisms from the production stage to the plants in the field. It allows to maintain higher concentrations of microorganisms and provides a temporary protective niche for microbial inoculants in the soil, either physically, via the provision of a protective surface of pore space (creating protective microhabitats) or nutritionally, via the provision of a specific substrate (Arora et al., 2010). The carrier constitutes the major portion (by volume and weight) of the inoculant (Malusá et al., 2012). A good carrier material must be (1) economical and readily available in adequate amounts, (2) easy to sterilize, (3) non-toxic to both microorganisms and plants, (4) have a high water retention capacity of more than 50%, (5) ensure the survival of the microorganisms and (6) have a sufficient shelf life (Herrmann & Lesueur, 2013; Sahu & Brahmaprakash, 2016) inoculant carriers can be broadly classified into five categories: soils, waste plant materials, inert materials, lyophilized microbial cultures, and liquid inoculants (Bashan et al., 2014).

Given the above factors, there is no single carrier that can offer all the desirable features, but good carriers should possess as many desirable features as possible. To produce inoculants, the target microorganism can be combined with a nonsterile or sterile carrier. From a purely microbiological perspective, a sterile carrier offers several advantages. However, sterilized carriers have drawbacks such as higher production costs, more labor requirements, the need for sterilizing equipment, and the requirement for aseptic packaging techniques.

The choice of carrier used in the manufacturing of inoculants is typically influenced by the method of application. There are two common categories of inoculants, those for direct application to the soil and those for seed treatment. These formulations can be in powdered form for seed treatment or granulated for soil application, due to the different means of delivery (Brahmaprakash & Sahu, 2012). When the major ingredient is alive and subject to changes, such as the case of microbial inoculants, it is important that the formulation remains stable during production, distribution, storage, and transportation to the farmer. Furthermore, good performance of an inoculant in the specific type of soil where it is applied is essential. Although tested inoculants come in a variety of forms, the commercial market offers only a limited number of inoculant varieties (Bashan et al., 2014).

11.1 Liquid inoculants

Liquid formulations can be broth cultures, suspensions in mineral or organic oils, oil-in-water suspensions, or polymer-based suspensions (Mahanty et al., 2017). The production of liquid inoculants involves a fermentation process, with the composition of the media and the culture conditions (i.e., temperature, pH, moisture content, agitation, and aeration) tailored to the specific strain (Brahmaprakash & Sahu, 2012). Liquid formulations can generally reach high population levels, allowing lower application rates while maintaining the same efficacy as other carriers. Besides, they are more attractive than solid inoculants due to their easier processing, longer shelf life of 1.5–2 years, and lower costs (Allouzi et al., 2022; Lobo et al., 2019). In addition, it is possible to add cell protectants and vitamins to the medium to enhance the yield and quality of the inoculant (Mahanty et al., 2017). The selection of additives employed in the production of liquid inoculants is based on their propensity to shield bacterial cells on seeds and in packaging from adverse environmental factors such as high temperatures, desiccation, and toxic conditions for seeds and seed compounds (Deaker et al., 2004). Many additives include high molecular weight polymers that are water-soluble, nontoxic, and have a complex chemical nature. These can be polymers such as polyethylene glycol (PEG), polyvinylpyrrolidone (PVP), gum arabic, sodium alginate, trehalose, and glycerol; adjuvants like carboxymethyl cellulose (CMC) and xanthan gum; and surfactants such as polysorbate (Brahmaprakash et al., 2020). Many studies have shown that the use of these additives enhances the resistance of applied microorganisms to abiotic conditions. For example, glycerol has a high capacity to protect cells from desiccation by reducing water loss rate, while trehalose is widely reported to improve cell tolerance to osmotic and temperature stress (Brahmaprakash & Sahu, 2012; Herrmann & Lesueur, 2013; Malusá et al., 2012). In general, maintaining a population higher than a 0.5% level of modification requires the addition of various osmolytes at a concentration of 1% or more.

A simple fermentation process can yield liquid inoculants, which can then be aseptically packaged directly from the fermenter and preserved. Minimizing the processing and sterilizing of the solid carrier material reduces production costs. With liquid formulations, thorough sterilization could be accomplished, and any contamination during storage could be quickly detected. Liquid inoculants offer convenience to farmers since they require less time, effort, and space to prepare than carrier-based formulations, and because they require less inoculum overall. The viable cell density of the required microorganisms, which essentially gives a suitable quantity of microorganisms on each seed, is the first yardstick to test the quality of the bioformulation. In addition, because liquid inoculants can be sprayed onto seeds as they move through seed drills and dry before reaching the seed bin on the planter, they are easily adapted to sophisticated seeding equipment.

11.2 Solid inoculants

Solid bioformulations are based on inorganic or organic carriers, prepared as granules or powders (Vassilev et al., 2020). These formulations can be prepared through solid-state fermentation, in which microorganisms grow on solid materials without the presence of free water, or by their mixing/coating on solid carriers (Vassilev et al., 2015). Solid-carrier-based bioformulation can be produced using soil materials (e.g., peat, coal, clays, inorganic soil), organic materials (e.g., compost, soybean meal, wheat bran, sawdust), or inert material (e.g., vermiculite, perlite, bentonite, silicates) (Herrmann & Lesueur, 2013). Furthermore, solid inoculants can be classified as either dry or wet formulations, depending on whether a drying method was used during production (Lobo et al., 2019). The most common types of solid formulations are dust, microgranules (MG), wettable powders (WP), granules (GR), and wettable/water-dispersible granules (WG, WDG) (Bashan et al., 2014; Guijarro et al., 2007). Binder, dispersant, wetting agents, and other ingredients are added to create them.

Dust is a very finely powdered mixture of the active component (~10%) with particle sizes ranging from 50 to 100 μm (Brahmaprakash & Sahu, 2012). This form of formulation has a long history of use but has encountered issues related to application and handling, despite occasionally being more effective at dispersing (Bashan et al., 2016). Moreover, granules are divided into two categories based on particle size: coarse particles (size range: 100–1000 μm) and microgranules (size range: 100–600 μm) (Arora et al., 2016). They consist of dry particles containing the active component, binder, and carrier. Granules typically contain 5 to 20% of their active components by concentration (Vassilev et al., 2020). The granules should be non-binder, non-dusty, and able to flow freely, while also breaking down into the active substance in the soil. They are usually applied in soil remediation and are considered safer because there is little risk of inhalation (Patil & Solanki, 2016). Storage and longer shelf life are particularly important considerations for granular formulations.

Another type of solid bioformulation is wettable, which includes wettable powders (WP) and wettable/water-dispersible granules (WG, WDG). WP are composed of 50%–80% technical powder, 15%–45% filler, 1%–10% dispersant, and 3%–5% surfactant by weight to achieve the desired potency formulation (measured in international units) and they are one of the oldest types of bioformulations (Fernandes Júnior et al., 2009). Despite being dry formulations, WP are easily soluble in water and can be added to a liquid carrier, usually water, right before application, which makes them of great interest. WG and WGD, also referred to as dry fluids, are nondusty, free-flowing granules that dissolve fast in the water, making them more user and environmentally friendly. Like those found in WP, they contain wetting agents and dispersion agents, with the dispersing agent typically present in higher concentrations. Both types of solid bioformulations exhibit excellent shelf life (Arora et al., 2016). For the preparation of powder formulations, industrial and agricultural materials waste by-products can be used, including organic cakes, cow dung–sand mixtures, sawdust–sand–molasses mixtures, corn cob–sand–molasses mixtures, bagasse–sand–molasses mixtures, inert charcoal, diatomaceous earth, and fly ash (Calvo et al., 2014).

11.3 Encapsulated cells

Microbe-based formulations created through encapsulation techniques are gaining popularity as a solution to the issues since they exhibit a variety of benefits over other solid and liquid formulations. Encapsulation of bacterial cells in polymers (e.g., alginate, polyacrylamide, pectin, and chitosan) is a technique to ensure the controlled release of the inoculum into soil maintaining their viability and effectiveness (Berninger et al., 2018). For encapsulation, the liquid inoculant is combined with an auxiliary polymer that can lead to solidification. The most widely used method is to mix the inoculum dropwise into a calcium chloride solution, resulting in the formation of solid beads with a high cell concentration (Sá et al., 2019). The encapsulation of PGPB provides protection to them against many various environmental stresses and enables their release into the soil, while the polymers are degraded by soil microorganisms (Bashan et al., 2016). The native soil microorganisms progressively dissolve the polymer, releasing the PGPB into the soil where plants in need of the inoculant are developing, thereby liberating the entrapped bacteria from the beads.

Water-soluble polymeric materials, including agar, methoxy-pectin, gellan gum, and mixtures of xanthan and locust bean gum, are also commonly used in the production of microbial-based products in bio-immobilization technology. However, alginate and carrageenan are the most used polymer-forming materials in microbial formulations to be introduced into soil–plant systems (Bashan et al., 2016; de-Bashan et al., 2012; Vassilev et al., 2020; Vejan et al., 2019); among these alginate is the preferred material for the majority of microbe encapsulations. Alginate is a natural anionic polymer and linear polysaccharide composed of 1,4′-linked β-D-mannuronic Acid and α-L-guluronic acid residues in different sequences, and is isolated from algae, *Pseudomonas* species, and *Azotobacter vinelandii* (Vejan et al., 2019). The primary advantages of alginate preparations are their nontoxicity, biodegradability, and gradual release of microorganisms into the soil (Bashan et al., 2002). This method has been successfully used to encapsulate plant-friendly bacteria such as *Azospirillum brasilense* and *Pseudomonas fluorescens*, which were later utilized to inoculate wheat plants in the field (Bacilio et al., 2017; Bashan et al., 2002, 2012; Galaviz et al., 2018; Gonzalez et al., 2018). These encapsulated bacteria were able to survive in the field for a sufficient amount of time, and their populations were on par with the longevity of bacteria derived from other carrier-based inoculants (Ahemad & Kibret, 2014). Furthermore, the addition of clay and skim milk to the beads significantly enhances bacterial survival compared to alginate beads alone.

Macro and microalginate beads are the two types of encapsulation technologies used. Macroalginate beads have a diameter size of approximately 1 to 4 mm, while microcapsules are from 50 to 200 μm in diameter or smaller. Alginate formulations have shown potential in addressing several of the issues associated with conventional peat inoculant (Bashan et al., 2016). These inoculant compositions could help overcome challenges in tropical low-input farming. After seeding and microbial inoculation, there is always a risk of prolonged dry periods in many tropical regions (Brahmaprakash & Sahu, 2012). Microorganisms are only released into the soil when enough moisture is available, which invariably coincides with the germination of seeds, while alginate-encapsulated formulations are already dehydrated owing to poor water activity.

The benefits and "know-how" of the immobilization technique used in the formulation of plant-beneficial microorganisms have been extensively described in several review publications. However, despite the clear advantages of controlled cell-release immobilized-cell formulations, their large-scale production and field application are currently limited. Since the price of the polymeric carrier is greater than that of the other solid and liquid formulation ingredients, one of the primary causes is the relatively high production cost (Bashan et al., 2014; Malusá et al., 2012; Vassilev et al., 2020).

There are numerous advantages and disadvantages of each of these types of microbial inoculants (Table 11.1). Each inoculant type has specific applications, but the same bioformulation can be applied to different crops and soil types. However, several factors must be considered, including the microorganism to be used, the intended purpose of application, the available infrastructure, and the required form of application.

Developing new formulations is challenging in microbiological practice, and shortcuts invariably lead to inoculant failure in the field. Formulation improvements are the key to the development of improved inoculants (Bashan et al., 2016). In recent years, there has been a growing interest in improving the quality of inoculants by using more sterile carriers than inoculants, as compared to nonsterile carriers that have performed poorly for many years. However, nonsterile carriers continue to be used due to the lack of formal standards in most countries involved in inoculant production (Bashan et al., 2014; Vejan et al., 2019).

A microorganism may function optimally under the care of a skillful crew and accurate laboratory conditions. However, formulating the inoculant into a user-friendly product that can be utilized by farmers without microbiological expertise, while still achieving comparable results under field conditions, is a challenging endeavor (Bashan et al., 2014). The effectiveness of an inoculant and farmer acceptation are often the most common barriers to commercialization.

Any formulation, particularly those containing live and sensitive ingredients, must exhibit stability during the production, distribution, storage, and transportation processes until it reaches the farmer, unlike farm chemicals (Bashan et al., 2016). To achieve this, quality control and standards must be established during every step. These standards should incorporate a series of parameters based on the existing knowledge about the efficiency of the product. Parameters such as media composition, temperature, pH, and aeration

TABLE 11.1 Advantages and disadvantages of different types of carriers.

Carrier	Advantages	Disadvantages
Liquid inoculants	Easy to handle and apply. Easy addition of additives to improve cell/spore production. Easy to produce and sterilize. High cell concentration, and low application rates. Compatible with modern agriculture types of machinery for its application. High shelf life (1–2 years)	Low viability during storage and on seeds. Cool temperatures for storage (4°C). More sensitive to stressful conditions. Susceptibility to contamination with other microorganisms
Solid inoculants	The raw material and production are lower. Easy transportation due to size. Easy to handle and apply. The inoculant can be applied by the farmer and is very convenient for the grower and therefore, popular.	Due to desiccation, the concentration of microorganisms is poor on average. Need a high concentration of cells in the development of the formulation. Bacteria will be in direct contact with pesticides applied to the seeds. Adhesion to the seed is poor, and much inoculant is lost during mixing and application.
Encapsulated cells	Provide temporary physical protection for the immobilized PGPB. The compounds used to prepare the encapsulates are nontoxic. Slow release of PGPB for successful root colonization. High cell concentration, and low application rates. Minimum storage space is required and can be dried at ambient temperature for prolonged periods.	The cost of some polymer compounds is expensive compared to peat, soil, and organic inoculants. Require furthermore expensive handling by the industry. More labor-intensive.

are some of the parameters that need to be continuously monitored and adjusted if necessary (Brahmaprakash & Sahu, 2012). Before the product release, cell identification (Gram stain), quantification of viable cells contained in the inoculum (commonly performed using plate count technique) and corroboration of the metabolic activity should be carried out.

When establishing a quality standard for bacterial inoculants, three important aspects must be considered: (1) maintain a minimum level of viable cells per unit, sufficient for plant inoculation and generating profits, (2) ensuring freedom from significant contamination, and (3) demonstrating consistent and reproducible efficacy under a range of field conditions (Brahmaprakash & Sahu, 2012). The literature describes many forms of tested inoculants (Table 11.2).

TABLE 11.2 Sample of formulation used as bacterial inoculants of different plant species in soil experiments.

Formulation	Presentation	Microorganism	Plant species/ use	References
None (culture media)	Liquid	*Bacillus megaterium, Bacillus paralicheniformis, Bacillus cabrialesii*	Wheat	Robles-Montoya, Chaparro-Encinas et al., (2020); Rojas-Padilla et al. (2020); Valenzuela-Aragon et al. (2019)
Carboxymethyl cellulose	Liquid	*Pseudomonas aeruginosa Bacillus licheniformis, Brevibacillus brevis, Micrococcus* sp.	Ryegrass *Jatropha curcas*	Viji et al. (2003); Jha and Saraf (2012)
Gum Arabic	Liquid	*Rhizobium* sp.	Rice	Wani et al. (2007)
Glycerol	Liquid	*Pseudomonas fluorescens*	Tomato	Manikandan et al. (2010)
Alginate	Solid wet formulation	*Azospirillum brasilense*	Wheat and tomato	Bashan et al., (2002)
Clay minerals, perlite/CMC	Solid wet and dry formulation	*B. japonicum, B. megaterium*	Soybean	Albareda et al. (2008)
Talc	Solid dry formulation	*P. fluorescens*	Chillies Mango	Bharathi et al. (2004); Commare et al. (2002)

TABLE 11.2 Sample of formulation used as bacterial inoculants of different plant species in soil experiments.—cont'd

Formulation	Presentation	Microorganism	Plant species/ use	References
Talc/ carboxymethyl cellulose (CMC)	Solid dry formulation	*Bacillus subtilis; Pseudomonas putida*	Rice	Saravanakumar et al. (2009)
Alginate	Encapsulation	*A. brasilense* and *Chlorella sorokiniana* *A. brasilense* *Bacillus pumilus* *Raoultella planticola*	Tomato and sorghum Desert trees Cotton	Galaviz et al. (2018); Trejo et al. (2012); Bashan et al. (2012); Wu et al. (2014)
Alginate and humic acid	Encapsulation	*B. megaterium* *Pseudomonas stuzeri*	Rice Peppers	Reetha et al. (2014) Bacilio et al. (2017)
CMC and corn starch	Encapsulation	*Herbaspirillum seropedicae; Gluconacetobacter diazotrophicus; Azospirillum amazonense;* and *Burkholderia tropica*	Sugarcane	da Silva et al. (2012)
Carrageenan	Encapsulation	*A. brasilense*	Guinea grass	Cortés-Patiño and Bonilla (2015)
Chitosan	Encapsulation	*Bacillus* strains	Tomato	Murphy et al. (2003)

A better understanding of the different conditions and characteristics of the interrelationship between soil, plant and microorganisms is necessary to improve the efficiency of bacterial inoculant application in the field. Distinct carrier materials for inoculation tests have been proposed in late decades, the type of formulation and its inoculation effectiveness depend on several environmental factors, such as the presence of other microbial species; water content, mineral nutrients, soil pH, and the species, genotype, and physiological state of the plant (Bashan et al., 2016; Malusá et al., 2012).

Despite the widespread use of both liquid and solid formulations in agrosystems, dry formulations are often chosen over wet formulations due to their longer shelf lives and ease of storage and transportation. The development of a bioformulation is proving to be an arduous task, and prior research in this area is insufficient. However, entrepreneurs are becoming more interested in this area, driven by the growing demand to develop new formulations to replace chemical pesticides and fertilizers. As a result, many projects are being supported to address this need.

Chapter 12

Performance evaluation of bacterial inoculants in the field

Marisol Ayala-Zepeda[1], Jonathan Rojas-Padilla[1], Alondra María Díaz-Rodríguez[1], Roel Alejandro Chávez-Luzanía[1], Fannie Isela Parra Cota[2] and Sergio de los Santos Villalobos[1]

[1]*Instituto Tecnológico de Sonora, Ciudad Obregón, Sonora, Mexico;* [2]*Campo Experimental Norman E. Borlaug, INIFAP, Ciudad Obregón, Sonora, Mexico*

Although in laboratory conditions the results of the inoculation of beneficial microorganisms have been satisfactory in many cases, demonstrating consistent and effective plant growth promotion and biocontrol effects, sometimes the application of the microbial inoculant in the field does not have the expected outcome (Bashan et al., 2014; Calvo et al., 2014; Khare & Arora, 2015). The reason for this variable impact is that the effectiveness of the microbial inoculant in the field is multifactorial and depends on (1) the target plant, (2) the microbial inoculant characteristics, (3) the management practices, and (4) the soil and environmental conditions (Fig. 12.1) and the interrelationship between those factors. Therefore, further analyses at the field level, considering a suitable active ingredient (microbial strain), the interaction between the microbial strain and the target crop, the edaphoclimatic

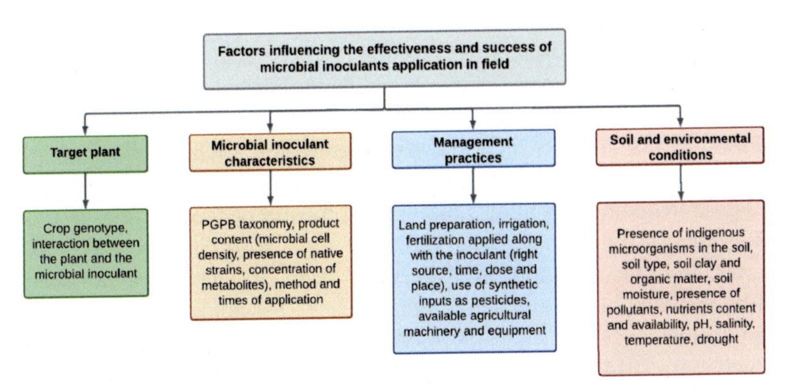

FIGURE 12.1 Factor influencing the effectiveness and success of microbial inoculants application in the field.

New Insights, Trends, and Challenges in the Development and Applications of Microbial Inoculants in Agriculture
https://doi.org/10.1016/B978-0-443-18855-8.00012-6

conditions, and management practices, must be conducted to accurately quantify the effects of the inoculant on crops (see Fig. 6.1(8)).

12.1 Microbial inoculant characteristics

When it comes to the efficacy of microbial inoculants, some traits of interest include the capability of the strains to survive in the holder and the soil, their ability to colonize the rhizosphere, their mobility in the ground (Grageda-Cabrera et al., 2018), and the differential production of various organic acids and other compounds (Sammauria et al., 2020). Many of these traits are regulated by the taxonomy of the strains, the microbial cell density or the concentration of metabolites in the product, the carrier, and the mode of application (Fig. 12.1).

In recent years, bacterial inoculant formulations and application methods have become more widespread, sophisticated, and more complex. To achieve better results, some formulations are customized according to the cultivation system, soil type, and the specific function or role that the microorganisms will have in the soil (Ahemad & Kibret, 2014; Kaushik & Djiwanti, 2019; Patil & Solanki, 2016). As discussed in the previous chapter, there are advantages and disadvantages associated with the different types of carriers (see Table 11.1).

The application methods also depend on the type of crop, soil, and agricultural practices/treatments in the field, and they also greatly influence inoculation efficiency. Direct soil inoculation is an easy and simple method of plant growth-promoting bacteria (PGPB) inoculation. Almost all types of bio-inoculants can be used for soil application by drenching, broadcasting, soil incorporation (mixed in the substrate or with the irrigation water), or micro-capsules (dos Santos et al., 2021). The most commonly used inoculants for direct application to the soil are liquid inoculants (liquid in nature or water-soluble formulations) and granular inoculants. These are applied after germination and go directly to the base of the plant, near to the plant roots, due to the limited movement of bacteria (Hou & Oluranti, 2013). In general, granular inoculants are placed in the furrow below or next to the seed, and in some cases, they could be applied by seeding machinery. This technique is used when a large population of a bacteria strain should be introduced to soil. Rojas-Padilla et al. (2022) applied microbeads of *Bacillus* strains (a PGPB) close to wheat seedlings, in the study, which showed a growth-promoting effect on plant phenotypic traits due to the increased exposure of the plants to the microbeads with PGPB. This method allows the gradual release of the bacteria, improves the adhesion, stability, and colonization of roots (dos Santos et al., 2021). However, soil inoculation avoids damage to seeds, the inhibition of the inoculant by the compounds applied to the seed (e.g., fungicides), and reduces the risk of losing inoculum when the seeds pass through the seeding machinery (Bashan et al., 2014; Rocha et al., 2019). In the case of liquid inoculants, their advantages are technical, such as the ease of handling for farmers.

Bioinoculants can also be inoculated by seed coating, where seeds are covered with a suspension of microorganisms or granules of bacterial inoculant, which allows delivery of PGPB to the rhizosphere of the target crop due to their establishment since germination. These techniques are a better option when increased germination is desired, as they confer protection to the seed against phytopathogens, and promote growth at the early stages of the plant (Moeinzadeh et al., 2010; Rocha et al., 2019). It is recommended to use an adherent because each seed needs to be covered with a considerable number of microorganisms. Most used adhesives are carboxymethyl cellulose (Viji et al., 2003), gum arabic (Wani et al., 2007), sucrose solution (Conga et al., 2009), and vegetable oils (Bashan et al., 2002). It is crucial to ensure that these adhesives are non-toxic for bacteria or seed. In addition, these adhesives can be used for both liquid and granular inoculants. A secondary role of the adhesive is to prevent the inoculant from decreasing during sowing, especially for powder inoculants when applied with air-seeders (Bashan et al., 2014).

Plant inoculation can be by foliar spraying or root dipping. Using a foliar spraying method, the plants can be inoculated with a high concentration of PGPB throughout the whole season or at certain growth stages, aiming to promote plant growth and fight plant diseases and pests at specific stages of the plant. However, similar to soil inoculation, it is sometimes recommended to use double amounts of the microbial inoculant compared to conventional fertilizers or pesticides to ensure their effectiveness (Preininger et al., 2018). Root dipping, on the other hand, is commonly for plantation crops such as trees, tomatoes, rice, onion, cole crops, and flowers. In this method, the seedling roots are dipped in a water suspension of the bioproduct for a sufficient period, therefore plant nursery preparation is required (Rocha et al., 2019; Thomas & Singh, 2019).

The design of microbial bioformulations and their delivery are challenging aspects and require an understanding of various factors, including modes of interaction, plant-root colonization, microbial adhesion to seeds, and antagonistic interactions (Zvinavashe et al., 2021). In this sense, to effectively deliver PGPB/PGPM, it is necessary to select the most suitable strain or microbial consortium for the desired impact on the target crop. Thereupon, an adequate formulation of the inoculant should be designed, followed by the selection of the appropriate application method. These latter two aspects control the desiccation, which can affect the viability of microorganisms (Fig. 12.2) (Zvinavashe et al., 2021).

12.2 Target plant

Plants play an essential role in the efficacy of the inoculant. Factors such as the crop genotype, physiological and nutritional status, and phenological stage rule the release of different root exudates and rhizodeposits. This, in turn, leads to the selection, inhibition, and stimulation of rhizosphere microorganisms' communities and their functions (Eligio Malusá et al., 2016).

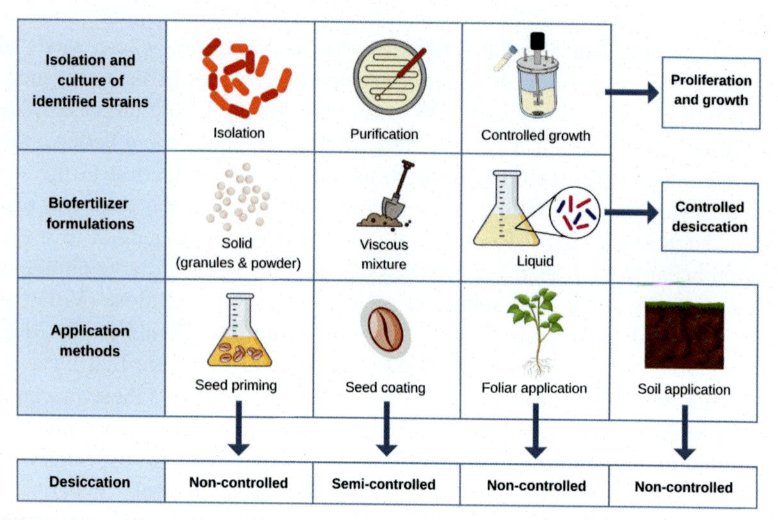

FIGURE 12.2 From the identification of the microorganism to the formulation and application a of microbial inoculant on the crop. The application method and the bioformulation characteristics play a crucial role in controlling the desiccation process (Zvinavashe et al., 2021).

For example, it has been highlighted that the application of microbial inoculants on seeds and seedlings can increase the efficacy of the inoculation, as root exudates are likely to be of great importance in initiating the rhizosphere effect in very young seedlings and on emerging lateral roots (Eligio Malusá et al., 2016).

In a field study, Di Salvo et al. (2018) compared three treatments: (1) control, (2) fertilized (kg N/ha), and (3) inoculated with two *Azospirillum brasilense* strains on wheat. Although the inoculation with *A. brasilense* strains and chemical fertilization with 46 kg N/ha did not modify the agronomic response of the wheat crop, they observed modifications in the genetic and functional characteristics of microbial communities associated with the rhizosphere of wheat under field conditions. This finding suggests that plant ontogeny mainly influences the physiology and genetic structure of rhizosphere microbial communities.

Similarly, de Salamone et al. (2010) conducted a study to investigate the eco-physiological response of rice plants to the inoculation with two *Azospirillum brasilense* strains under field conditions. They analyzed the rhizosphere bacterial community of the inoculated rice plants and isolated diazotrophic bacteria with biological nitrogen fixation (BNF) and their physiology. Additionally, the genetic diversity of the endophytic bacterial communities was characterized in different parts of rice plants. *A. brasilense* inoculation increased biomass and nitrogen accumulation in the plants, but did not contribute to nitrogen derived from BNF. This suggests that the inoculation of *Azospirillum* had a hormonal effect and induced physiological responses in the plant tissues, improving the efficiency of nitrogen absorption and resulting in superior yields.

Correspondingly, the success of the method used in the field to inoculate beneficial microorganisms depends on the crucial relationship that exists between the host plant and the microbial inoculant association. This association can occur on different levels of interaction and relies on a fine-tuned adjustment of metabolism from both organisms. Thus, the genetic characteristics of both the microorganisms and the host plant can define the level of the inoculation response during their association (de Salamone et al., 2010; Silveira et al., 2016).

12.3 Management practices

The use of synthetic fertilizers, herbicides, insecticides, and fungicides, as well as practices like tillage, crop rotation, and other agronomical practices that depend on the farmer, can have an impact on the inoculation effect. Management practices such as soil preparation, irrigation, and fertilization also influence the growth and activity of the inoculant (Compant et al., 2019; Kamilova et al., 2015).

Fertilization is known to be the agronomic practice that most significantly affects the efficacy of microbial inoculants (Eligio Malusá et al., 2016). Glick (2012) exemplifies, as reported in the literature, that a bacterium that promotes plant growth by providing fixed nitrogen or phosphorus is unlikely to provide the same benefit to crops when high amounts of chemical fertilizer are applied to the soil. For example, Galindo et al. (2019) demonstrated that lower doses of nitrogen fertilization in conjunction with the inoculation of *A. brasilense* in field trials, enhance the efficiency of nitrogen fertilization and increase the yields of wheat crops.

Several studies have demonstrated that the effect of the inoculation varies according to the total nitrogen content in the soil, particularly the nitrogen added through fertilization, as well as the residual (initial) nitrogen content. In this regard, management practices are a core component that influence and modulate soil characteristics and, consequently, the effectiveness of the inoculation.

Benavides et al. (2019) evaluated two tillage systems, conventional tillage and minimum tillage, in sweet corn crops under sustainable management practices, and established the following treatments: (1) control without fertilization or inoculation; (2) inoculation with Nitragin Maize AZ39 *Azospirillum brasilense*); (3) fertilization with 75 kg N/ha + inoculation; and, (4) fertilization with 150 kg N/ha. The study found that the fertilized treatment with 75 kg N/ha + inoculation, along with a minimum tillage system, would be a suitable alternative. This treatment obtained the same yield as the fully fertilized treatment. In addition, the use of the minimum tillage system was found to promote soil resource conservation.

Fungicides can have an impact on various stages of the symbiosis process, from the survival of the strains on the seed to their mechanisms and metabolism, such as nodule formation and N_2 fixation efficiency in the case of rhizobia. Nevertheless, the detrimental effects on microbial inoculants become more pronounced when multiple chemicals (fungicides, insecticides, and herbicides) are combined. This is a common practice in agricultural lands to facilitate the combined control of weeds, pests, and diseases (Santos et al., 2021). The adverse effects of some pesticides on microbial inoculants, from the moment of contact to the damage caused to the plant are summarized in Fig. 12.3.

On the contrary, some pesticides can be compatible with the inoculation of certain microbial strains. Cardozo et al. (2022) conducted a study over three consecutive seasons (2015–2018) to assess the effects of foliar inoculation of *Azospirillum brasilense* Az39 in maize crops with different combinations of herbicides (glyphosate and atrazine). In the treatment with foliar inoculation of Az39, along with one application of atrazine and two applications of glyphosate (considered a suitable combination of herbicides based cost and environmental risk), higher yields and chlorophyll content were observed.

In general, the application of microbial inoculants must not require specialized equipment to increase the acceptance and success among farmers. Opting for simple methods and traditional machinery is considered the preferable approach (Malusá et al., 2012).

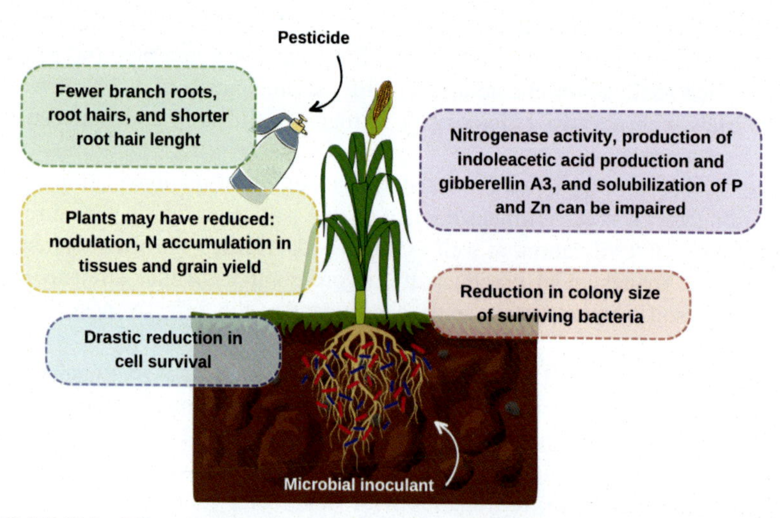

FIGURE 12.3 Effects reported on the incompatibility between some pesticides and inoculants, from the moment of contact with microbial cells to the damage to plant development (Santos et al., 2021).

12.4 Soil and environmental conditions

The effectiveness of microbial inoculant is influenced by various factors, including temperature, humidity, pH, and soil chemical components such as N, P, Ca, S, Mg, Mo, Fe, and Co content, which impact the population dynamics of any introduced microbial species (Grageda-Cabrera et al., 2018).

The physiological, morphological, and metabolic responses of soil microorganisms, as well as the availability of soil nutrients, are highly influenced by soil pH. Within a pH range of 5–7, the majority of micronutrients (B, Cu, Fe, Mn, Ni, and Zn) are more readily available, while macronutrients (N, K, Ca, Mg, and S) are more available within a pH range of 6.5–8. However, P availability is highest within a pH range of 6–7. Outside of these optimal ranges, nutrients become less available to microbes and plants (Haby 1993; cited by Khare & Arora, 2015). For example, it has been reported that acidic pH reduces the activity of many soil organisms (Khare & Arora, 2015).

Thilakarathna and Raizada (2017) conducted trials comparing different rhizobia inoculants in field-grown soybean and observed that, under moderate soil pH conditions (pH 6.6–7.8), the inoculants led to the highest increases in nodule number compared to the corresponding uninoculated controls. However, this trait showed a significant decline when the soil was more acidic or alkaline. The study also highlighted that the availability of soil nitrogen concentrations played a crucial role in the survival, abundance, and diversity of soybean rhizobia in soil, as well as their nodulating activity.

Similarly, factors such as aeration and moisture content are crucial for ensuring optimal microbial activity; characteristics that soil in good physical condition (including favorable soil texture) have. Thus, pore size distribution actively determines the fate of introduced microorganisms, and variability in the performance of bacteria in different textured soils may be related to differences in the available pore spaces. Hence, soil texture significantly influences various soil processes (Khare & Arora, 2015).

Recently, the use of chemical compounds labeled as ^{15}N isotope techniques has been used in conjunction with microbial inoculants (Adesemoye et al., 2010) to assess the effectiveness of bioinoculant in the field. Labeled compounds, such as ^{15}N enriched fertilizers, are commonly employed as tracers to track their fate in a complex system like the soil (Barraclough, 1995). Some authors recommend ^{15}N-based studies to quantify the N-fertilizer uptake alongside with the inoculation of PGPM/PGPB on crops (Kennedy et al., 2008). These studies provide insights into the actual impact of the inoculated strains on crops in terms of nitrogen use efficiency.

On the other side, conducting studies on water use efficiency (WUE) under different agronomic practices using both isotopic techniques (e.g., keeling plots, isotopic mass balance) and conventional methods (e.g., soil moisture sensors, microlysimeters, eddy covariance, modeling on Aquacrop—developed by the Food and Agriculture Organization of the United Nations, FAO—) are strongly recommended. These studies should be carried out in conjunction with the application of microbial inoculants to design and improve sustainable management practices, considering sowing dates, fertilization, irrigation, and current and future conditions.

In the study by Benavides et al. (2019), they compared two tillage systems on sweet corn (previously described) and evaluated the nitrogen use efficiency (measured using urea enriched with 2% atom excess of ^{15}N) and water use efficiency (measured using a neutron probe). The fertilized treatment with 75 kg N/ha along with inoculation showed higher nitrogen use efficiency compared to the treatment that received 150 kg N/ha without inoculation. Regarding water use efficiency, no significant difference was observed between the tillage systems. However, the fertilized and inoculated treatments demonstrated greater efficiency in water use compared to the unfertilized and uninoculated control.

In addition, Martins et al. (2018), inoculated the strains (1) *Azospirillum brasilense* Sp245, (2) *A. brasilense* AbV5 + AbV6, (3) *Herbaspirillum seropedicae* ZAE94, on maize (*Zea mays* L.), and had (4) an inoculated control. In this study, the *A. brasilense*-based inoculants showed improved the ^{15}N-urea acquisition efficiency and enhanced grain quality and yield compared to the uninoculated control.

In a multi-year field trial conducted by Kumar et al. (2007), spanning three years of field trials, significant increases in yield and nitrogen, phosphorus, and potassium (NPK) content in the root and stem of maize were observed due to soil inoculation with *Pseudomonas corrugata*. Similarly, Rinu and Pandey (2009) reported the positive influence of *Bacillus subtilis* inoculation on plant biomass and yield-related parameters on lentils in a 2- year field experiment (Rinu & Pandey, 2009).

The inoculation of maize seeds with *Azotobacter chroococcum* and *Azospirillum brasilense* resulted in statistically significant improvements in plant yield and nutrient content, as reported by Pandey et al. (1998). Similary, the inoculation of *Pseudomonas fluorescens* strain Pf-102, Pf-103, and *Acinetobacter rhizosphaerae* BHIB 723 in pea seeds increased various parameters including germination percentage, root and stem length, vigor index, dry weight, and yield (Gulati et al., 2009; Negi et al., 2005).

Other factors that influence the effectiveness of microbial inoculants include the competition between the inoculated microorganisms and the natural soil flora, as well as the presence of various root exudates and pollutants (Sammauria et al., 2020). In addition, it has been reported that pathogens and beneficial microorganisms modulate similarly early defense genes, some of which are related to a shift in the nitrogen metabolism to control the microbial colonization through the production of certain compounds (Bilgin et al., 2010).

de Salamone et al. (2010) conducted a study on the inoculation of *Azospirillum brasilense* on rice plants and found that the competence of *Azospirillum* to establish itself in the paddy soil with rice should have been very high due to the presence of a specifically adapted microflora in the rhizosphere of this plant. However, the inoculation of *A. brasilense* had a clear effect on community structure, modifying the density functionality, and composition of the rhizosphere communities of the rice plants under field conditions.

In this sense, these bioproducts have proven to contribute to the conservation of soil and microbial diversity by enhancing the beneficial edaphic microbiota. They also enhance crop resource efficiency—water and nutrients— while simultaneously improving the efficiency of chemical fertilizers (Fan et al., 2015; Kumar et al., 2016).

It is widely accepted that the efficiency of growth promotion mechanisms, such as those described in Chapter 5, is influenced by the ability of the inoculum to colonize the medium (Chávez-Luzanía et al., 2022; Rilling et al., 2018). Such colonization and the expression of beneficial plant functions are affected by the plant species, phenological stage of the crop, defense mechanisms, seed hydrophobicity, presence of chemical compounds on the seed surface, ambient temperature, the radiation received, rainfall, humidity, agricultural soil management practices, type of fertilization, crop rotation, physical and chemical properties of the soil, presence of organic matter, and soil microorganisms (Rilling et al., 2018).

The main challenges in evaluating the effect of field-applied microorganisms are that some are very closely related to each other and can only be detected by detailed analysis of their 16S rRNA gene or other genes within the species (Borriss, 2011). For example, *Bacillus* are commonly studied and marketed as plant growth promoters for many crops, but strains of this genus, such as *Bacillus subtilis* and *Bacillus amyloquefaciens can exhibit* PGPB or saprophytic characteristics, or both (Calvo et al., 2014; Tumbarski et al., 2014). Thus, colonization with one PGPB species may be concealed by saprophytic rhizosphere inhabitants of the other species (Posada et al., 2016).

Therefore, it is necessary to determine whether the inoculum used is capable of establishing itself and expressing its functions, or not, and to determine its population and how it fluctuates in order to draw conclusions about its effect. However, the complexity of the soil−plant matrix represents a critical point in obtaining reliable data on the population of the microorganism of interest under field conditions (Manfredini et al., 2021). In the last two - decades, research efforts have intensified in the search of methods capable of determining the presence and quantifying the population of specific microorganisms under field conditions (Manfredini et al., 2021; Rilling et al., 2018). Among these, we can find the following methods: (1) molecular, (2) immunological, and (3) based on reporter genes (Fig. 12.4) (Romano et al., 2020).

Employing a molecular method, Chávez-Luzanía et al. (2022) designed specific primer pairs for a wheat plant growth-promoting bacterial consortium

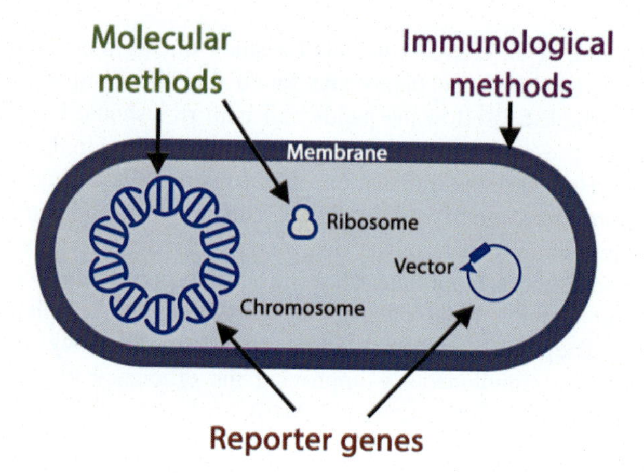

FIGURE 12.4 Types of methods for monitoring microorganisms and the cell structures in which they act.

composed of *Bacillus cabrialesii* TE3^{T}, *Bacillus paralicheniformis* TRQ65, and *Priestia megaterium* TRQ8. Through a pangenomic approach and utilizing genomic databases, unique regions in the genomes of the microorganisms of interest were identified. Based on these regions, pairs of primers were designed that hybridize specifically with each strain. This approach enables the detection of each microorganism in agricultural soil samples at concentrations of 1×10^5 cells per gram of soil, which allows monitoring of the presence of the consortium under field conditions.

On the other hand, Mourya & Jauhri (2002) developed a method based on the use of the TnS-*lac*Z reporter gene for the detection of *Pseudomonas striata* P-27 in the soybean rhizosphere. To implement this method, the reporter gene was inserted in *P. striata* P-27 through transposition using *Escherichia coli* S17-1. The transformed microorganisms were then used for inoculating soybean plants. Their presence was assessed by the observation of color change in Petri dishes using serial dilutions, which allowed the determination of populations of up to 1×10^3 CFU per gram of soil at various phenological stages of the crop.

Krishnen et al. (2011) employed the immunoblot method to monitor the presence of *Pseudomonas fluorescens*, *Azospirillum brasilense*, and *Rhizobium leguminosarum* in the rice crop. This methodology allowed the quantification of the inoculum up to 147 days of cultivation, enabling the establishment of a correlation between changes in irrigation events and the population dynamics of the inoculum. Such insights into the population dynamics were instrumental in understanding the impact on the performance of the inoculum.

To summarize, the effect of the microbial inoculants is influenced by the interdependence of various factors. These factors include the characteristics of the microbial inoculant, such as strain taxonomy, cell density and concentration, bioformulation, delivery method, times of application, and survival; the target crop's characteristics, such as genotype, interaction with the microorganisms, phenological stage, nutritional requirements, and its exudates; the management practices, including tillage methods, machinery and equipment used for land preparation, NPK fertilization, irrigation system, and the use of pesticides and other inputs; and the edaphoclimatic conditions, such as soil type, structure, pH, moisture, nutrient content, presence of pollutants, salinity, temperature, CO_2 concentration, and presence of indigenous microorganisms, which contribute to the overall effect.

A proper characterization of the target soil and rhizospheres, as habitats for introduced microorganisms, is key to the development of bioformulations (Khare & Arora, 2015). However, only a limited fraction of the numerous studies published in the literature on the inoculation of plants with PGPB/PGPM have discussed delivery methods (Bashan et al., 2016). In this sense, each agricultural land must be considered as an individual case of study in order to establish appropriate and ideal recommendation for microbial inoculant application and agricultural management that are realistic and attainable for the farmer.

The aforementioned considerations would maximize the potential of microbial inoculants and their beneficial impact on crops in terms of increases in crop productivity, yield, and quality, as well as seeking the reduction of the environmental impact that conventional fertilization and management practices—as the indiscriminate use of pesticides, or inefficient irrigation systems—cause, contributing to soil restoration, the conservation of microbial diversity, reduction of economic losses, and the ensurance of agroecosystem services. In addition, the integration of various scientific fields, including microbiology, material sciences, and biotechnology, among others, will create innovative and practical bioproducts, leading to significant advancements in sustainable agroecology.

Part III

Present and future scenario

Chapter 13

The role of microbial culture collections in the development of microbial inoculants

Ramón Arteaga Garibay, Lorena Jacqueline Gómez-Godínez, José Martín Ruvalcaba Gómez and Hugo Alberto Zaldivar-López
Centro Nacional de Recursos Genéticos, INIFAP, Tepatitlán de Morelos, Jalisco, Mexico

Microorganisms have long been valued as vital factors of world biodiversity. They play a major part in soil fertility, industry, and health and are employed in diagnostics and efficacy testing of medicines, biocides, detergents, or reference strains. This versatility has brought microorganisms to the fore, turning them into crucial players in the formation of the biotechnology period in recent times. Despite its importance, preserving microorganisms has not yet been given due importance. Exorbitant amounts of microorganisms have been isolated from natural populations for a long time; however, only a very small amount has been preserved, while many of them have been neglected and even lost (Gold, 1992; Whitman et al., 1998).

Several methods aimed at the preservation of microorganisms have been developed, and each of them exhibits its advantages and disadvantages. The most common preservation strategies include cold storage, freeze-drying, and preservation in liquid nitrogen. Cold storage is the simplest and cheapest method, but it is not suitable for all strains of microorganisms. Freeze-drying and storage in liquid nitrogen are more advanced methods that provide greater stability and long-term preservation (Berninger et al., 2018).

The objective of the conservation of microorganisms is to ensure the genetic pool as a resource for the continuous search for new products that can favor different areas such as medicines, nutrition, and agriculture, among others. As industries develop and improve in various biotechnological processes, they discover microorganisms and their functions. Therefore, it is necessary to increase the number of conservation protocols to ensure the maintenance of at least three aspects of the microbial isolates: integrity, purity, and functionality; aimed to achieve their use in screening, reproduction, characterization, and sustainable production (Sharma et al., 2017).

New Insights, Trends, and Challenges in the Development and Applications of Microbial Inoculants in Agriculture
https://doi.org/10.1016/B978-0-443-18855-8.00013-8

For that reason, the microorganism conservation process implemented by the culture collections plays an extremely important role in the safeguarding of microbial resources to allow their sustainable use and serve as providers, in the short, medium, and long term, of authenticated biological material for both research and teaching purposes, with high quality and commitment in the form of reference strains, necessary for quality control (Díaz-Rodríguez et al., 2021; Smith et al., 2014).

13.1 Establishment of culture collections

Several publications about techniques and procedures for microbial conservation are available, including some guidelines for the establishment of culture collections. Nonetheless, no internationally approved set of guidelines, covering all aspects of culture collection activities, had been compiled until the World Federation for Culture Collections (WFCC) prepared such a document in 1991: The WFCC Guidelines for the Establishment and Operation of Collections of Cultures of Microorganisms. These guidelines were prepared to help those culture collections that offer services outside of their host institutions; and focused on the following aspects: organization, funding, objectives, the type and the number of strains to be maintained, staff, services, safety, and quality standards (Sharma et al., 2017).

The WFCC is the international coordinating organization for culture collections since The International Union of Biological Sciences and the International Union of Microbiological Societies recognized it in 1970. A series of handbooks describing the resources available to culture collections have been initiated by WFCC; volumes concerned with filamentous bacteria, fungi, and yeasts (Hawksworth & Kirsop, 1988; Kirsop & DaSilva, 1988; Kirsop & Kurtzman, 1988).

The World Data Center for Microorganisms (WDCM), now managed by the WFCC, produced the first World Directory of Collections of Cultures of Microorganisms in 1972, and it has been frequently updated ever since. Nowadays, this information is supplemented by the Microbial Strain Data Network (MSDN), established in 1985. All information is supported and curated by the Committee on Data for Science and Technology, the WFCC, and the International Union of Microbiological Societies. Likewise, data related to the strains by individual collections may be consulted online (Kirsop, 1988; Rogosa et al., 1986; Staines et al., 1986).

Depending on the scope and objectives, culture collections are classified into four categories.

1. **Service collections**, that serve as verified-cultures suppliers. This type of collection, and the offered materials, are usually listed in public access catalogs. Regularly, the government is the main funder of these collections, and supply can be directly or indirectly supported. Biological material in these types of collections is generally given in exchange for recovery costs.

2. **In-house collections** are usually established to attend to the individual requirements of organizations, institutions, or companies. Catalogs of this type of collection are generally unavailable, or access is restricted; hence, cultures are supplied on a discretionary basis, but collections of this type can be very substantial.
3. **Research collections** are mainly built by individual scientists as part of their research programs. Such collections are often unique because they include novel and unusual strains in restricted groups but long-term storage facilities are rarely adequate, and resources allow cultures to be available only to close colleagues. When scientists change positions, retire, or pursue different lines of research, this type of collection is often lost.
4. **Laboratory suppliers** are generally dedicated to providing single strains of species used in teaching or research. Prices offered are usually below those that service collections must pay.

The funding arrangements for the maintenance of culture collections vary greatly, but apart from some service collections sponsored by national governments or by international coalitions of collections, funds are rarely adequate for the long-term to enable collections to maximize their value as a resource to the scientific community.

13.2 Global perspective of culture collections

The culture collections have a long history. Prof. Frantisek Karl of Prague could be regarded as the pioneer in the establishment of culture collections as we know them nowadays. He was the first one to collect microbial cultures and made them available to others for a fee. Today, culture collections have a variety of roles, with the emphasis on cultures being for taxonomy and epidemiology, and their number has also increased significantly. Currently, there are 831 culture collections registered in the WDCM database on February 10, 2023 (https://ccinfo.wdcm.org/statistics), distributed across 78 countries (Table 13.1). These culture collections are supported by different government and non-government agencies, and currently safeguard a total of 3,370,507 microorganisms preserved, including bacteria, fungi, viruses, and cell lines. The optimal infrastructure and expertise in the preservation, characterization, documentation, and identification of every group of microorganisms are crucial for the fulfillment of the mission of culture collections (OMPI, 2017; Federación Mundial de Colecciones de Cultivo (FELACC), 2010; WDCM, 2022).

13.3 Services offered by microbial culture collections

Microbial culture collections are important for different reasons related to their scope, including research development, industry support, and food safety, among many others. Their primary roles include collecting, maintaining, and

TABLE 13.1 Examples of microbial resource centers (microbial collections) around the world.

Culture collection	Acronym	Country	Website	IDA[a]
Agricultural Research Service Culture Collection	NRRL	United States of America	http://nrrl.ncaur.usda.gov/	Yes
Coleccion Española de cepas tipo	CECT	Spain	https://www.uv.es/uvweb/coleccion-espanola-cultivos-tipo/es/coleccion-espanola-cultivos-tipo-1285872233521.html	Yes
Collection Nationale de Cultures de Microorganismes	CNCM	France	https://research.pasteur.fr/en/team/national-collection-of-cultures-of-microorganisms	Yes
American Type Culture Collection	ATCC	United States of America	https://www.atcc.org	Yes
Coleccion de microorgansimos del CNRG	CM-CNRG	México	http://cmcnrg.inifap.gob.mx/acerca.html	Yes
The National Collection of Type Cultures (NCTC) for bacteria	NCTC	United Kingdom	https://www.phe-culturecollections.org.uk/collections/nctc.aspx	Yes
NITE Biological Resource Center	NBRC	Japan	https://www.nite.go.jp/en/nbrc/index.html	Yes
Colección de Microorganismos Edáficos y Endófitos Nativos	COLMENA	México	http://www.itson.mx/colmena	Yes
The National Measurement Institute	NMI	Australia	http://www.measurement.gov.au	
Leibniz-Institut DSMZ - Deutsche Sammlung von Mikroorganismen und Zellkulturen GmbH (DSMZ)	DSMZ	Germany	https://www.dsmz.de/	Yes
Bioresource Collection and Research Center	BCRC	Taiwan	http://www.bcrc.firdi.org.tw	
Collections Coordonnées Marocaines de Microorganismes	CCMM	Morocco	https://www.cnrst.ma/fr/	Yes

[a]*IDA: International depositary authority under the Budapest treaty.*

dispatching microbial cultures, and making them accessible to the microbiological research community through the publication of printed or electronic catalogs. In the end, the different users should be able to access the data of cultures because they are as important as the organisms to achieve the sustainability of the collections (Sharma & Shouche, 2014). Efficiency during the transfer of knowledge and information between the collection and its users depends to a great extent on advanced databases, printed catalogs, or online databases (Díaz-Rodríguez et al., 2021). At this point, researchers and taxonomists can choose the strains for their potential application. In the bioinformatics era, these databases will become even more valuable since microbial collections act as safe deposits of microorganisms with restricted distribution, as well as providing identification services according to the expertise of the culture collection about the different types of microorganisms.

Depositing cultures described in scientific publications and used in research contributes to ensuring their availability for future access and would allow the reproducibility of research protocols. In addition, to stimulate the formation of highly qualified human resources, culture collections develop training programs, specialized courses, and workshops related to the identification and maintenance of microorganisms; therefore, it is essential for this purpose, to deposit, along with the microbial strains, the associated derived sequences into the public repositories. These training strategies are essential to training personnel from medical, environmental, industry, or government laboratories who have responsibility for the isolation and identification of microorganisms, disease diagnosis, quality control, fermentation, culture management, etc. (Anand et al., 2022; De Vero et al., 2019).

One of the more realistic indicators of the efficacy of the current system of culture collections is the number of species that they preserve, especially the number of those that have a known potential use or benefit, that represents a possible economic benefit. Microorganisms do not have a history of inventory production on a regional basis, unlike birds, mammals, and arthropods. Only rarely is there an attempt to study all the different habitats available to a single group of organisms due to the intensiveness required for sampling, culturing, and massively identifying microorganisms (Becker et al., 2019).

13.4 Microbial genetic resources for agriculture

Microorganisms are also referred to as Microbial Genetic Resources (MGRs), which include algae, bacteria (including cyanobacteria), fungi (including yeasts), and certain protistan groups. MGRs are usually studied in laboratories rather than institutions, botanic gardens, or germplasm banks. For this reason, they do not get much attention in overall reviews of biological diversity and global genetic resources (Gilmour et al., 2022; Shuli Liu et al., 2022; Montaño López et al., 2022).

Despite the wide commercial exploitation of MGRs, they remain disproportionate in comparison to the key roles that microorganisms play in the biosphere. The genetic resources movement has tended to emphasize the importance of microorganisms, perhaps linked to the common misconception that they do not merit consideration. However, in so many ecosystems, microorganisms play some important roles, such as symbionts (endophytes, mycorrhizae, and in insect guts), biological nitrogen fixers (rhizobia, cyanobacteria, cyanobacteria-containing lichens), biodegrading agents of dead animal and plant material, and as organisms for population control of plants and insects through natural biocontrol (Pham et al., 2019).

The aim of microbial culture collections is not only to achieve long-term viability but also to maintain the genotypic and phenotypic stability of its preserved cultures since that integrity is essential for the authentication of previous findings. Researchers often claim that the culture collection provides a mutant version of the expected wild-type strain. This indicates that culture collections may be dealing with mutants rather than original wild microorganisms. Therefore, a comprehensive characterization of cultures on morphological, anatomical, physiological, immunological, and molecular grounds is a must before and after preservation (Maske et al., 2021; Wu et al., 2017).

In 1974, the United Nations Educational, Scientific, and Cultural Organization (UNESCO) started an inventory of microbial resources centers (MIRCENs). The infrastructure for an international network geared to the management, distribution, and utilization of the world's microbial gene pool is expected to be provided by some MIRCENs with two main functions: (1) conservation of agroecosystems through the isolation and preservation of the microbial diversity, providing a biological safeguard service; and (2) to facilitate the study and efficient access of microorganisms, reference strains, and microbial resources by the public to generate biotechnological strategies. Culture collections have adopted innovative and cutting-edge technologies to add value to the offered services, as the main role of MIRCENs was initially for their value in carrying out both taxonomic and epidemiological studies (Sharma et al., 2017).

The role of MIRCENs as supporters for agriculture activities related to microorganisms is essentially to serve as primary suppliers of culturable microorganisms, replicable parts of these DNA, genomes, extrachromosomal material, and viable but not yet culturable microorganisms in biological or environmental matrices (Anand et al., 2022; Díaz-Rodríguez et al., 2021).

It is important to highlight that there are specific requirements to maintain and ensure the quality of services for MIRCENs (Fig. 13.1) (Sharma & Shouche, 2014).

(1) **Microbial culture characterization and authentication.** The purpose of this service is to achieve accurate strain identification as well as preservation. Stable maintenance of the preserved organisms should be ensured, as should the survival of at least 70% of the preserved cells.

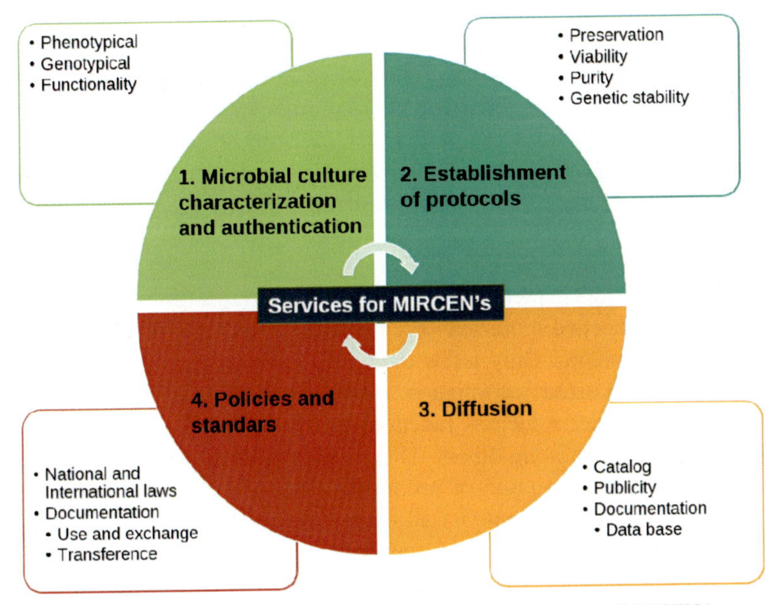

FIGURE 13.1 Services provide for microbial resource centers (MIRCENs).

(2) **Establishment and validation of protocols** focused on the standardization of the culture collections operation, which includes activities from the preservation of different types of microorganisms to the maintenance of specialized equipment. For this purpose, it is necessary to have highly trained staff in cutting-edge technologies for the preservation, growth, and identification of microorganisms.

(3) **Diffusion.** Achieving a greater scope and accessibility of the generated information to the public is paramount to achieving the sustainability of culture collections (catalogs and documentation).

(4) **Policies and standards.** Compliance with national and international laws, regulations, and policies about safety, shipping, and exchange of microorganisms.

The services offered by most of the MIRCENs around the world are limited to the preservation, deposit, and transfer of microorganisms and do not thoroughly study the strain capabilities; usually, this is left to interested researchers or professionals related to the agriculture field with limited information.

13.5 The role of microbial culture collections in the development of microbial inoculants

Maintaining certain groups of microorganisms as regular soil dwellers is good for maintaining balance in the environment. Microbials with indigenous potentials are the most appropriate alternative to agrochemicals to reduce the negative impact associated with their use without demeriting on the productive potential of the soil. To ensure a beneficial effect on plants, microorganisms demand to be cultivated under optimal conditions that allow them to express their functions. Agriculturally important microorganisms include those that are mainly involved in nitrogen fixation, the transformation of organic matter, nutrient cycling, formation and aeration of the soil, and plant growth regulation (Iriti et al., 2019; Ney et al., 2020; Robles-Montoya, Chaparro-Encinas et al., 2020).

The research and characterization of promising strains for biotechnological applications are needed more than ever, as well as making publicly available the related information. Only a few culture collections have a public catalog containing systematized information about preserved microorganisms that allow search by criteria such as function, biosafety, taxonomy, or biotechnological applications, among others. This lack of publicly accessible information complicates the selection of promising strains and delays the design of biotechnological developments based on the use of microorganisms, such as their use as inoculants in agricultural production (Becker et al., 2019). For this reason, MIRCENs must also make the biological data of microorganisms accessible and include data about the source of isolation (biological or environmental context), evolutive characteristics, taxonomy, metabolic characteristics, and ecological relationship by the creation of the database (online platform) with open access to easier exchange and transfer of microbial strains (Díaz-Rodríguez et al., 2021).

That level of development, in the context of culture collections, can only be achieved by directing resources toward fundamental services and involving expertise related to four crucial aspects of the development of microbial inoculants (Fig. 13.2): the (1) specific ex situ preservation and quality assurance; (2) identification, authentication, and taxonomic classification; (3) data management and sharing; although some collections provide identity and accession numbers to microorganisms, only a few microbial culture collections provide information on traits of microorganisms in the public domain. And, (4) support for the implementation of the applicable legislation.

On the other hand, molecular tools and techniques such as next-generation sequencing (NGS) technologies, RNA seq, and transcriptomic and metagenomic approaches have exponentially increased the identification and detection of both culturable and nonculturable microorganisms. This advancement has allowed for a more accurate knowledge of the biodiversity of microorganisms associated with particular niches (Paterson et al., 2017). It is noteworthy that NGS technologies have become more affordable in terms of costs while offering increasingly high-quality results in less time. Therefore, it is expected that the identification of novel species will be based on their entire genome in the future. Sequences of correctly described species can be used as

FIGURE 13.2 Support of microbial culture collections. (A) Preservation, (B) identification and taxonomic classification, (C) data management, and (D) support for legislation.

a reference to infer lineages of so-far-uncultured microbial species in natural populations.

Most of the culture collections only accept cultures that have been properly characterized and accompanied by the supporting documentation necessary for its traceability for assigning accession numbers (De Vero et al., 2019). These collections usually have the infrastructure, personnel, and equipment to reverify the culture's identity before the accession numbers' assignation and they employ long-term storage protocols for preservation. Culture collections have adopted new techniques based on molecular biology to add value to their services. The advancement of biochemical and physiological studies focused on the study and characterization of microorganisms has greatly improved the efficiency and effectiveness of the preservation of organisms.

The use of standardized formats for providing information and the implementation of state-of-the-art programming interfaces would allow true and effective interoperability of collection information systems with bioinformatics databases and software. In some cases, the microbial collections have not been organized, highlighting the need to assign attributes to microorganisms beyond identity and accession number. The functionally well-equipped culture collections generally have enough staff and funds to sustain their operations over time, but some collections lack the necessary resources for smooth functioning. While most of the microbial culture collections across the globe preserve microorganisms mainly to supply cultures for research purposes in academic and research institutions, there is also increasing commercial use of such microbes in the industry (De Vero et al., 2019).

Nowadays, biodiversity management and the sharing of bioresources are regulated by the country of origin, following the provisions of the Nagoya Protocol. For this reason, the use of guidelines of the World Federation for Culture Collections (WFCC) is highly recommendable for some of the collections, although some of them are functional and work for both research and commercial purposes. Those guidelines are available on the official website of the WFCC (http://www.cabri.org/guidelines/micro-organisms/M100Ap1.html). Alternatively, the guidelines developed by the OECD can also serve as a reference (https://www.oecd.org/science/emerging-tech/23547773.pdf) for establishing microbial culture collection worldwide (Sharma et al., 2019).

The World Data Centre for Microorganisms (WDCM) of the World Federation for Culture Collections (WFCC) manages the global catalog of microorganisms (GCM), which combines as many online catalogs as possible, creating a robust system that links catalog entries with all types of data and makes them accessible through a single portal. The WDCM acts as the establishment of policies and rules for the transfer and sustainable use of preserved microbial genetic resources for the development of microbial inoculants, which are related to the fair and equitable sharing of the benefits derived from the use of genetic resources according to the Nagoya protocol. Regulation applicable to culture collections and the use of biological resources worldwide started with the entry into force of the Convention on Biological Diversity (CBD) on December 29, 1993, followed by the Nagoya Protocol (NP) on Access and Benefit Sharing (ABS) on October 12, 2014. Culture collections proactively serve as benchmarks for the design of solutions for the agri-food sector to implement the CBD and its NP. MOSAICC, the Micro-Organisms Sustainable use and Access regulation International Code of Conduct, and TRUST, the TRansparent User-friendly System of Transfer, are forerunning initiatives concerning the implementation of the NP. The development of GCM is the breakthrough in the global handling of the NP (Desmeth, 2017), acting as an information broker between the online catalog and the users. Information related to the possession, location, transfers, and use of microbial strains, as well as the country of origin, creation of derived patents, and all associated scientific publications can be retrieved by the system. GCM is a transparent and sustainable data management system that can be used to build safe, ethical, and balanced ABS processes at a global level. The role of culture collections between providers and users of microbiological resources has been strengthened by GCM (Wu et al., 2017).

Finally, the Budapest Treaty regulates the process of patent deposits in an international framework; thus, a patent deposit made with one International Depositary Authority is sufficient and recognized by all member states of Budapest. Non-member countries may also accept deposits according to the Budapest Treaty, following their own country's norms. In both cases, microbial collections has an important role in the management of microbial genetic resources, particularly in the agriculture industry on the development of microbial inoculants, due to its importance in basic activities related to management, use and exchange, and exploitation under policies for sustainable use of these natural resources (Bussas et al., 2017).

The loss of microbial biodiversity must be opposed and prevented from both an ecological and a socioeconomic point of view, as they play a fundamental role in maintaining natural ecosystems and in developing valuable applications in the agro and food industries (de los Santos-Villalobos et al., 2021). The debate on the management of global diversity is largely ignored due to a generalized misperception of the microbial world. The naked eye cannot see the microorganisms on the other side. Microbial collection has a vital role to play in addressing societal challenges. The quality and credibility of science are ensured by guaranteeing the distribution of high-quality microbial resources for the development of microbial inoculants. In the future, they will be in charge of the preservation of uncultivable organisms and must face the challenges involved in safeguarding the resources to be delivered to future generations.

In conclusion, microbial culture collections play a critical role in the development of microbial inoculants. They provide a repository for microbial diversity and a source of microorganisms that can be explored for their potential as inoculant candidates. Microbial culture collections can also be used to screen for microorganisms with specific traits and provide reference strains for quality control. These collections contain well-characterized and typified strains that can be used as standards to ensure the consistency and effectiveness of bioformulations. Microbial culture collections are essential for promoting sustainable agriculture by providing access to microorganisms that can enhance plant growth, increase nutrient availability, and suppress plant pathogens.

Chapter 14

Legal framework for the development of microbial inoculants

Alejandra Miranda Carrazco[1], Alondra María Díaz-Rodríguez[2], Fannie Isela Parra Cota[3] and Sergio de los Santos Villalobos[2]
[1]*Departamento de Ciencias Ambientales, UAM Unidad Lerma (UAML), Lerma Estado de México, Mexico;* [2]*Instituto Tecnológico de Sonora, Ciudad Obregón, Sonora, Mexico;* [3]*Campo Experimental Norman E. Borlaug, INIFAP, Ciudad Obregón, Sonora, Mexico*

Microorganisms have been used conscientiously in agriculture for many years, including organic products containing beneficial microorganisms (such as composted substrates) and more recently, pure and mixed cultures selected for their capabilities for plant-growth promotion, biocontrol, and soil health improvers, leading to a formal market for biological-based agricultural amendments.

The bioinoculants industry (formulation, commercialization, and use) has increased remarkably in the last years worldwide. The environmental issues caused by anthropogenic activities, the excessive use of synthetic fertilizers, soil erosion, and the impetus provided by the demand for safe and healthy food have led the search for sustainable alternatives in agriculture to replace synthetic inputs. Moreover, the substantial rise in global food demand due to population growth, urbanization reducing land available for agriculture, and the expanding global economy with increased per capita annual income pose challenges to improve crop yield and quality (FAO, 2018). Thus, the development of biological products has reached a level where they can compete against chemical fertilizers due to their environmental and economic advantages. Eventually, biological-based products could potentially partially or totally replace conventional agricultural products (Barquero et al., 2019).

Many nations, such as the United States (US) and those in the European Union (EU), have been investing a substantial amount in scientific research in the organic fertilization area (Bharti & Suryavanshi, 2021).

New Insights, Trends, and Challenges in the Development and Applications of Microbial Inoculants in Agriculture
https://doi.org/10.1016/B978-0-443-18855-8.00014-X

Legal regulation is necessary to guarantee the quality of microbial inoculants, safeguard farmers and consumers' rights, and guide the commercial promotion of microbial products (Fig. 14.1). However, regulation of microbial inoculants represents a challenge around the world. Biological agricultural products are still largely regulated by chemical fertilizer laws due to the lack of experience with microorganism-based products (Arjjumend & Koutouki, 2021).

The term "microbial inoculant" can include several categories: microbial fertilizers, biofertilizers, biopesticides, biostimulants, and organic fertilizers (Fig. 14.2). The absence of a consensus on the definition of these products is the first issue to address in establishing a framework for the development of microbial inoculants. Moreover, the lack of communication between the

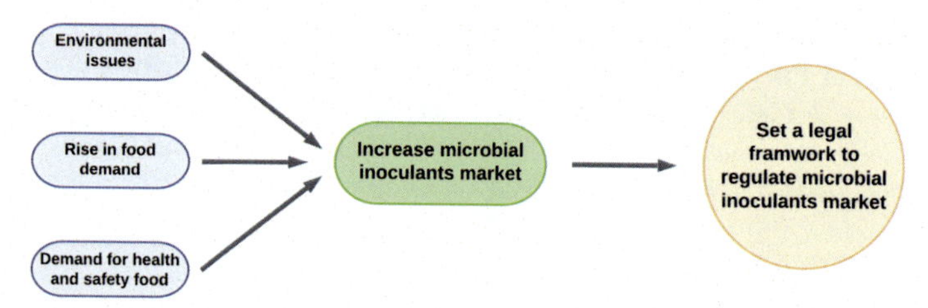

FIGURE 14.1 The factors that demand to set a legal framework to regulate the bioinoculants market.

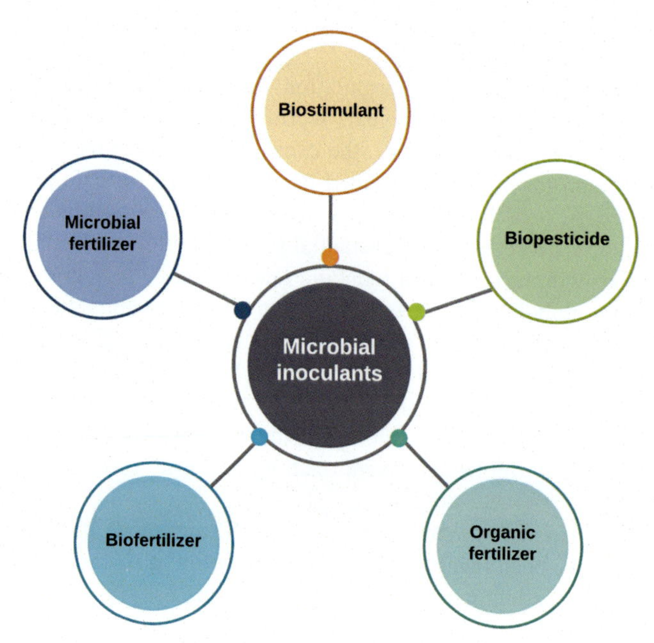

FIGURE 14.2 Types of bioinoculants.

scientific community and farmers has hindered the legal regularization that meets agricultural yield, sustainable practices, and human health.

As the most important component in microbial inoculants is the microorganisms, the microbial quality of the products must be reported, including identification, abundance, plant growth-promoting capabilities, and contamination level. Moreover, physical and chemical factors, mode of application, dosage, etc. (Fig. 14.3) are aspects that affect the quality of the product, and therefore, they should be known by the end user (Bharti & Suryavanshi, 2021).

Currently, there are many organizations around the world that aim to regularize the production, commercialization, and use of organic inputs in agriculture. Particularly, the EU has made significant changes in the fertilizer legal framework to include biostimulants. In their most recent document, Regulation 2019/1009, which lays down the rules on safety, quality, and labeling requirements for fertilizing products, they defined plant biostimulant as "a product stimulating plant nutrition processes independently of the product's nutrient content with the sole aim of improving one or more of the following characteristics of the plant or the plant rhizosphere: (a) nutrient use efficiency, (b) tolerance to abiotic stress; (c) quality traits, (d) availability of confined nutrients in soil or rhizosphere". Microbial and non-microbial plant biostimulants are included in this definition. Microbial amendments must fulfill several requirements to be on the market, including the (1) name of microorganisms; (2) taxonomic classification of the microorganisms (levels genus, species, and strain); (3) data background in scientific literature reporting safe production, conservation, and use of the microorganisms; (4) taxonomic relation to microorganism species fulfilling the requirements for a Qualified

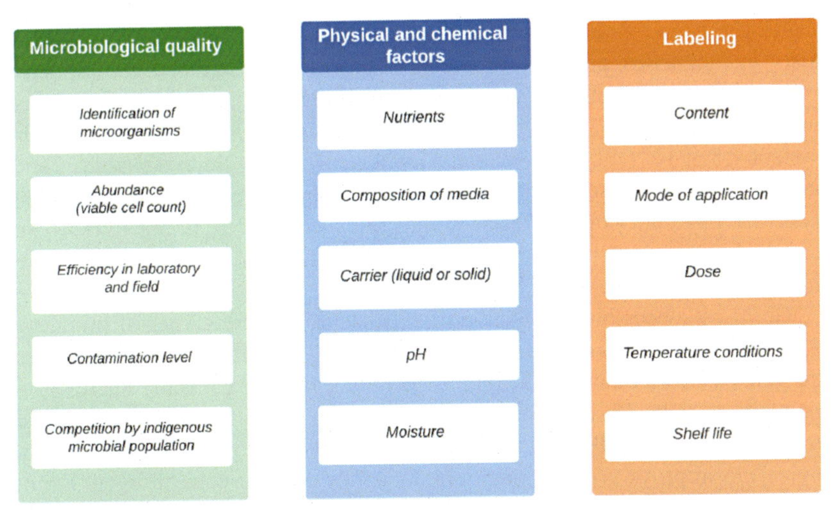

FIGURE 14.3 Quality factors of bioinculants that must be reported by producers.

Presumption of Safety as established by the European Food Safety Authority; (5) information of the production process; (6) information on the identity and residue levels of residual intermediates, toxins or microbial metabolites in the component material; and the (7) natural occurrence, survival, and mobility in the environment. Moreover, the possible contamination contained in organic fertilizers is delimited. The microorganisms considered pathogens are *Salmonella* spp. (absence in 25 g or 25 mL) and *Escherichia coli* or *Enterococcaceae* (maximum 1000 in 1 g or 1 mL), *Listeria monocytogenes* (absence in 25 g or 25 mL), *Vibrio* spp. (absence in 25 g or 25 mL), *Shigella* spp. (absence in 25 g or 25 mL), *Staphylococcus aureus* (absence 25 g in 25 mL), *Enterococcaceae* (10 CFU/g), anaerobic plate counts unless the microbial plant biostimulant in an anaerobic bacterium (10^5 CFU/g or mL), and yeast and mold counts unless the microbial plant biostimulant is a fungus (1000 CFU/g or mL).

In addition, some European countries have their own regulations. Spain has a Royal Decree 999/2017 that considers some aspects of microorganisms included in fertilizers. According to this decree, only organisms that have been shown to stimulate plant development could be used. To commercialize fertilizer products containing microorganisms, they must be identified according to 16S rRNA and 18S rRNA gene sequences for prokaryotes and eukaryotes, respectively. Moreover, the description of isolation and quantification of the microorganisms from the fertilizer, the growth conditions of the microorganisms, and the genetic material isolation, as well as the PCR conditions to amplify 16S rRNA or 18S rRNA, must be provided. Also, agricultural efficiency must be reported.

Otherwise, Italy's Legislative Decree 75/2010 includes the following products as biostimulants: protein hydrolysate of medicinal plants; hydrolysate from the animal epithelium (solid or liquid); liquid extract from medicinal plants, algae, and molasses, and acid extract from algae from the fucal family. Also, encompasses products based on mycorrhizal fungi and rhizospheric bacteria. Requirements are (1) endomycorrhizal fungi capable of forming punctual entries into the roots of the host plants (30%), and (2) bacteria content $<5 \times 10^6$ CFU (only *Pseudomonas fluorescens*, *Bacillus subtilis*, and *Streptomyces* sp. are allowed). The biostimulant action must not be derived from the phytohormonal effects of the product. Also, it is not allowed to declare biostimulant properties to product mixtures mentioned above with other kinds of fertilizers.

Germany places biostimulants in its legislation in three types of products: plant strengtheners, soil improvers, and plant aid agents. Plant strengtheners are defined as substances and mixes, including microorganisms, intended to maintain plant health and protect plants from nonparasitic impairments. Soil improvers aim to influence soil properties (chemical, physical, and/or biological) to improve growth conditions for crops and promote nitrogen assimilation. Finally, plant aid agents are substances that act on plants to achieve a benefit in the plant structure, production technique, or use technique. However, there are no special requirements for microbial preparations.

Austria regulates the marketing of fertilizers through the Austrian Federal Office for Food Safety, particularly by the Federal Law on the trade and other fertilizer products, which was set in 2021. It does not have a special section for biostimulants or any product related to microorganisms. However, it considers farm manure, soil improvers, and plant aid agents among fertilizer products. The Law of Bio-fertilizers of Montenegro defines a fertilizer as a "chemical compound of mineral and organic origin and a mixture of these compounds, regardless of the physical state, as well as certain microorganisms, which are used for direct or indirect fertilizing and improvement of the land fertility". The allowed fertilizer types are mineral (inorganic) fertilizers, organic fertilizers, and microbiological fertilizers. Particularly, organic fertilizers are products that consist of a minimum of 50% of organic matter and at least 1% of nitrogen, 1% of phosphor, and 1% of potassium. Microbiological fertilizers are fertilizers containing certain useful microorganisms in terms of plant growth promotion. Biostimulators are considered fertilizers products and are defined as "substances of synthetic or natural origin (hormones, vitamins, amino acids, humus acids, etc.) that have a stimulating effect on physiological biochemical processes in plants".

Otherwise, China has experienced an amazing increase in agriculture since economic reforms in 1978 reached family-based farming (Fukase & Martin, 2016), leading to investment in agricultural research and development. Nowadays, China is one of the leading countries in promoting sustainable agriculture. For instance, biofertilizers were included in the "Agricultural Biological Products Development Action Plan" in 2015, and the Ministry of Agriculture introduced the "Zero Growth of Chemical Fertilizer and Pesticides Use" to eliminate the excessive application of synthetic fertilizer and reduce greenhouse emissions. Moreover, China has set several standards on strain safety, product specification, terminology, labeling, testing methods, and technical specifications for inspection and evaluation (Ruan et al., 2020).

China defines biofertilizer/microbial fertilizers as a type of fertilizer containing living microorganisms with specific functions (Chinese Agricultural Standard NYT 1113-2006). They recognize several microbial inoculants: *Rhizobium* inoculant, *Azotobacteria* inoculant, inoculant of phosphate-solubilizing microorganisms, silicate bacteria inoculant, inoculant of photosynthetic bacteria, organic matter-decomposing inoculant, inoculant of plant growth-promoting rhizosphere microorganisms, multiple species inoculant, mycorrhizal fungi inoculant, bioremediating inoculant (Ruan et al., 2020). The Chinese government implemented registration management on biofertilizers in 1996, and since that moment, the regulation has changed according to the farmer's necessities and scientific advances to identify and characterize the microorganisms used as microbial inoculants. Biofertilizer evaluation in China includes microbial identification and toxicological tests of functional microorganisms and end-product. The identification must be done by a national authoritative institution unless the strain is coming from a national culture collection institute, so the identification test can be exempted (Fang, 2020). Moreover, some parameters must be measured: (1) appearance, color, odor, and physical state; (2) amount of living target bacteria, which depends on the bacterial phenotype (fast-growing *Rhizobium*, slow-growing *Rhizobium*, N fixation bacteria, Si and P mineralizing bacteria and multi-strain biofertilizer); (3) water content; (4) size; (5) organic matter, which must be >20 in powder and granular products; (6) pH of 5.5–7.0 for solid and 6.0–7.5 for powder and granular products; (7) percent of non-target bacteria contamination (< 20%); and (8) valid period (> 6 months) (Suh et al., 2006).

The fertilizer (control) order (FCO) (No. 11–3/83-STU, 1985) was released in India in 1985 to regulate the fertilizers market (sale, price, and quality), but it was until 2009 that biofertilizers were added. The Indian Ministry of Agriculture defines a biofertilizer as “the product containing carrier-based (solid or liquid) living microorganisms which are agriculturally useful in terms of nitrogen fixation, phosphorus solubilization or nutrient mobilization, to increase the productivity of the soil and/or crop”. They considered 10 categories of biofertilizers: *Rhizobium*, *Azotobacter*, *Azospirillum*, phosphate solubilizing bacteria, mycorrhizal biofertilizers, potassium mobilizing biofertilizers, zinc solubilizing biofertilizers, *Acetobacter*, carrier-based consortia, and liquid consortia (Arjjumend & Koutouki, 2021). Moreover, algal biofertilizers are also available in the market, but they have not yet been approved by the Fertilizer Control Order (Khurana & Kumar, 2022). To market a biofertilizer in India, producers must register the following specifications: (1) carrier base; (2) viable cell count, $> 10^7$ CFU/g or mL; (3) contamination level, which must not have contamination at a 10^5 dilution; (4) pH; (5) particle size; (6) moisture percent, 30%–40%; and (7) efficiency character. Every parameter depends on the type of biofertilizer, for instance, the efficiency character of every type of biofertilizer is listed in Table 14.1.

Also, organic fertilizers are considered in the same FCO, they are defined as “substances made up of one or more unprocessed material(s) of a biological nature (plant or animal) and may include unprocessed mineral materials that have been altered through microbiological decomposition process”. Besides, the Union Ministry of Agriculture and Farmer’s Welfare (2021) regulates microbial inoculants, and it defined them as “substance of microorganisms or a combination of both whose primary function when applied to plants, seeds or the rhizosphere is to stimulate physiological processes in plants and to enhance its nutrient content”. The products included in this category are botanical extracts, vitamins, protein hydrolysates, amino acids, cell-free microbial products, antioxidants, antitranspirants, humic and fulvic acids, and biochemicals.

In Canada, fertilizers (plant nutrients) and supplements (products other than fertilizers that improve the physical condition of the soil or aid plant

TABLE 14.1 Efficiency requirements for bio-fertilizers in India according to the fertilizer (control) order (FCO) (No. 11–3/83-STU).

Type of biofertilizer	Efficiency character
Rhizobium	Should show effective nodulation on all the species listed on the packet.
Azotobacter	The strain should be capable of fixing at least 10 mg of nitrogen per g of sucrose consumed.
Azospirillum	Formation of a white pellicle in semisolid N-free bromothymol blue media.
Phosphate solubilizing bacteria	The strain should have phosphate solubilizing capacity in the range of a minimum of 30% when tested spectrophotometrically. In terms of zone formation, a minimum 5 mm solubilizing zone in prescribed media has at least 3 mm thickness.
Mycorrhizal bio-fertilizers	The infectivity potential must be 80 infection points in test roots/gm of mycorrhizal inoculum used.
Potassium mobilizing bio-fertilizers	Maximum 10 mm solubilization zone in prescribed media having at least 3 mm thickness.
Zinc solubilizing bio-fertilizers	Maximum 10 mm solubilization zone in prescribed media having at least 3 mm thickness.
Acetobacter	Formulation of a yellowish pellicle in semisolid medium N free medium.
Carrier-based consortia	The efficiency character depends on the consortia members, and the criteria of every type of biofertilizer are applied.
Liquid consortia	The efficiency character depends on the consortia members, and the criteria of every type of biofertilizer are applied.

growth and crop yield) are regulated by the Canadian Food Inspection Agency (CFIA) under the Fertilizers Act and Fertilizers Regulations. These regulations include fertilizers, micronutrients, lawn and garden products, and supplements (water-holding polymers, microbial inoculants, abiotic stress protectants, liming materials, and waste-derived materials such as composts and municipal biosolids). The CFIA oversees product safety and labeling. Particularly, microbial supplements involve pure cultures of bacteria or fungi, microbial consortia, and genetically modified microorganisms (including microorganisms generated through gene editing or synthetic biology techniques). To market a microbial supplement, the following aspects must be reported: identification of active microorganisms (except in the case of consortia) by genus and species (and strain if available), a minimum number of viable cells per gram, or another description of the minimum concentration for microorganisms that are not viable cells. For microbial consortia, identification is not necessary, but the descriptor of the concentration of viable microorganisms is required.

In March 2019, the United States Environmental Protection Agency (EPA) released the draft "Guidance for Plant Regulator Products and Claims, including Plant Biostimulants", which defines plant biostimulants as "products containing naturally occurring substances and microbes that are used to stimulate plant growth, enhance resistance to plant pests, and reduce abiotic stress. They can promote greater water and nutrient use efficiency but are not intended to provide any nutritionally relevant fertilizer benefit to the plant". However, there is no federal regulation for microbial inoculants, but each state can establish several requirements for their use and commercialization. In some states, biofertilizers must be tested for efficacy, while in other states, it is only necessary to notify that the product is on the market (Kamilova et al., 2015).

Latin American countries are top consumers of biofertilizers (Malusá & Vassilev, 2014). Some countries are leading in the field of microbial inoculant regulation. The Southern Common Market (MERCOSUR), a union of countries from South America (Argentina, Brazil, Paraguay, Uruguay, Venezuela, and Bolivia), released the resolution "Recommendations on the use of inoculants in Agriculture" (MERCOSUR/GMC/RES N 28/98) in 1998, which establishes some regulations about beneficial microbes' commercialization between countries. Some requirements are (1) the use of strains recommended for competent bodies, (2) products must be free of contaminants and be formulated on sterile support, and (3) a period of validity > 6 months, among others. Also, the normative sets the official methods for inoculant analysis (Kamilova et al., 2015).

Otherwise, Chile enacted law 21349 in 2021, which establishes the rules regarding the composition, labeling, and commercialization of fertilizers and biostimulants. It defines a biostimulant as a "substance o mix of substances or microorganisms, applicable to seeds, plants, or rhizosphere, that stimulate natural processes of plant nutrition, to improve the efficiency in the nutrient use, abiotic stress tolerance, quality attributes, or the availability of immobilized nutrients in soil or rhizosphere". The label of the products must contain their composition, contaminants, and quality controls. The Agricultural and Livestock Service is in charge of regulating the procedure of sampling and analysis for the composition verification and quality parameters.

Mexico does not have specified norms for the microbial inoculant market, despite the rapid growth of its industry. The registration of microbial inoculants is currently regulated by the Federal Commission for Protection against Health Risks (COFEPRIS); however, all microbial inoculants registered with this entity end up classified between pesticides or plant nutrients due to the lack of specific regulations for these products. In terms of biological effectiveness, in Mexico, the Secretariat of Agriculture, Livestock, Rural Development, Fisheries and Food (SAGARPA), through the National Service of Health, Safety and Agrifood Quality (SENASICA), is responsible for establishing the requirements and specifications for conducting biological effectiveness studies of plant nutrition inputs and agricultural pesticides, and their technical opinion through the Mexican Official Standards NOM-077-FITO-2000 and NOM-032-FITO-1995, respectively. However, it is essential to update these standards for the technologies based on microorganisms. Moreover, the NOM-182-SSA1-2010 regulates the labeling of plant nutrients, for the case of inoculants, it states that the product must indicate the genus and species of the microorganisms contained in the product. Currently, a draft standard is being developed for the use and commercialization of biofertilizers within the country. This law is expected to promote the gradual replacement of inorganic fertilizers with biofertilizers.

Despite the many approaches around the world for microbial inoculant regulation, a greater effort is necessary to define the concept of microbial inoculants and their derivates, and homogenize standards regarding content, efficiency, contamination level, carriers, storage conditions, physical and chemical properties, validity, etc., for their use and commercialization (Malusá & Vassilev, 2014). Cooperation between scientists, producers, sellers, and farmers is crucial to set microbial inoculant legislation.

Chapter 15

Perspectives and challenges for innovating microbial inoculants

Alondra María Díaz-Rodríguez[1], Fannie Isela Parra Cota[2] and Sergio de los Santos Villalobos[1]
[1]*Instituto Tecnológico de Sonora, Ciudad Obregón, Sonora, Mexico;* [2]*Campo Experimental Norman E. Borlaug, INIFAP, Ciudad Obregón, Sonora, Mexico*

There is no more urgent time than now to intensify global research efforts to understand the complex relationships established between plants and microorganisms and to generate sustainable bio-agrological products to achieve healthier, more resistant plants with better crop yields. The recent and ongoing research, as well as the technological progress, provide a unique approach to describing the intricate interactions between plants and microorganisms. Our knowledge of the underlying mechanisms of plant–microorganism interactions and how they influence the plant's health and growth is still developing, as technology opens new paths for researchers to answer the questions that will lead to sustainable agriculture.

For example, new techniques used in molecular research have greatly enhanced our comprehension of the regulation of plants' defense mechanisms against pathogens. Gene silencing has proven very useful to elucidate the molecular pathways involved in plant defense and resistance, pathogen and PGPM recognition by the plant, pathogen production of toxins and their virulence, and PGPM production of antibiotics and siderophores (Muhammad et al., 2019).

Research in recent decades has shown that the use of PGPM and BCA-based inoculants represents an alternative to improve plant nutrition, stress tolerance, control phytopathogenic diseases, and mitigate environmental problems generated by the use of conventional synthetic fertilizers. At present, the international market for these bioproducts has been valued at over US $2.3 billion in 2023. It is also expected to reach US $4.09 billion by 2028, with an annual growth rate of 12.1%, between 2023 and 2028 (Mordor Intelligence, 2023). A wide range of microorganisms has proven to have beneficial effects on plants, either benefiting nutrition, growth, yield, water intake, indirect and/or direct pathogen control, and abiotic stress relief. Unfortunately, there is not

New Insights, Trends, and Challenges in the Development and Applications of Microbial Inoculants in Agriculture
https://doi.org/10.1016/B978-0-443-18855-8.00015-1

a single universal microbial inoculant that works for all crops or conditions, in comparison to chemical fertilizers like urea; however, their large number of mechanisms, their specificity, and their results represent an advantage over conventional ones (Brahmaprakash & Sahu, 2012). Most studies have been carried out, with great success, in vitro or in vivo at a small or medium scale, but their beneficial results are rarely reproduced in the field as expected.

The determination of whether the inoculant and its specific mechanism by which it was selected are indeed responsible for the effects of promotion and biocontrol produced in the plant is still a field that needs to be explored in more detail; but, untangle the contribution of each PGPM to plant growth/biocontrol effects and clarify the potentially complex interactions with other microorganisms in the soil is a huge challenge. Future research must consider other factors in plant–microorganism relationships, such as interaction with the entire microbiome and the environment. Ecological, metagenomic, population genetics, metabolomic, transcriptomic, and other kinds of studies are necessary for a holistic understanding of the complex network of interactions that constantly occur in proximity to plants (Kong & Liu, 2022). This will bring us closer to an explanation of the variations in the results in the field and help us improve efficiencies in the application of microbial inoculants.

In addition, it has been shown that the use of purified bioactive compounds, such as phytohormones, antibiotics, or solubilizing agents produced by microorganisms themselves, can have an important plant growth-promoting or biocontrol effect. However, research related to the interaction between bacteria, fungi, and purified compounds is necessary. Although positive improvements when combined are superior to those obtained separately, it is necessary to explore these effects in the field to generate useful agricultural tools for current and especially future needs. Using what we know and what we are learning about the plant–microorganism interactions to create strategies for the control of plant diseases, enhance plant growth, make better use of soil nutrients, and create more resistant and better-yielding crops is still a major challenge for current and future research.

In summary, the success of microbial inoculant development depends on innovative strategies related to (1) the metabolic-functional and molecular identification of PGPM, (2) their formulation, and (3) their correct application in the field. These three points require collaborative efforts from multidisciplinary experts in soil chemistry, biotechnology, microbiology, and agronomy. Together they can create and provide highly useful, safe, easy-to-apply, and economically acceptable biotech products that will serve as environmental solutions and important inputs in sustainable agriculture.

However, there are main challenges in the development of these bioproducts, such as (1) the large-scale production of inoculants, (2) accessibility of inoculants for farmers of different socioeconomic classes, (3) minimization of the variability of field results, and (4) the lack of legal requirements and controls to improve the quality of inoculants. Once these challenges are solved, this technology will achieve an increase in economic profits, due to the reduction of synthetic agricultural inputs and the increase in crop yields.

The cost of production is an important constraint in the development of microbial inoculants, considering that their price should not exceed that of conventional ones to ensure the sustainability of the market. For wide adoption by farmers, an inoculant must be cost-effective, compatible with standard machinery, and combinable with traditional techniques such as seed treatments and phytosanitary control (Berninger et al., 2018; Malusá et al., 2012). It is necessary to carry out socioeconomic impact studies in different regions worldwide where the use of this technology, which is not so recent, has been favored. On the other hand, it is relevant to address the advantages in the sustainable economy of using these bioproducts from the context of the bio-economy and the implications it would have in emerging communities (Matson, 2012).

The maximum integration of scientific methods with social participation is necessary from all sectors for the generation, acceptance, legislation, and execution of this type of alternative. A constraint limiting the use of microbial inoculants may be the lack of awareness or reliability among farmers. This could be targeted with the proper transfer of knowledge about the technology and providing instructions of their application in workshops/training and demonstration trials so that producers can directly address all their concerns (Odoh et al., 2020). With this interaction between scientists and farmers, it will be possible the development of improved microbial inoculants tailored to the needs of each farmer, increasing their confidence and addoption.

On the other hand, efforts must be aimed to develop bioformulations with a greater diversity of carriers and strains with different metabolic potentials, as well as diversifying the forms of application of the product to expand its use in all crops and different agricultural systems. Besides, extensive research is needed to study the effectiveness of inoculants in different crops while considering climate conditions and agricultural practice factors, such as fertilizer inputs and tillage practices. These efforts would increase approval and higher application by farmers.

References

Aasfar, A., Bargaz, A., Yaakoubi, K., Hilali, A., Bennis, I., Zeroual, Y., & Meftah Kadmiri, I. (2021). Nitrogen fixing azotobacter species as potential soil biological enhancers for crop nutrition and yield stability. *Frontiers in Microbiology, 12*(February), 1–19. https://doi.org/10.3389/fmicb.2021.628379

Abadi, V. A. J. M., Sepehri, M., Rahmani, H. A., Zarei, M., Ronaghi, A., Taghavi, S. M., & Shamshiripour, M. (2020). Role of dominant phyllosphere bacteria with plant growth–promoting characteristics on growth and nutrition of maize (*Zea mays* L.). *Journal of Soil Science and Plant Nutrition, 20*(4), 2348–2363. https://doi.org/10.1007/s42729-020-00302-1

Ádám, A. L., Nagy, Z., Kátay, G., Mergenthaler, E., & Viczián, O. (2018). Signals of systemic immunity in plants: Progress and open questions. *International Journal of Molecular Sciences, 19*(4), 1–21. https://doi.org/10.3390/ijms19041146

Adeniji, A. A., Babalola, O. O., & Loots, D. T. (2020). Metabolomic applications for understanding complex tripartite plant-microbes interactions: Strategies and perspectives. *Biotechnology Reports, 25*, e00425. https://doi.org/10.1016/j.btre.2020.e00425

Adesemoye, A. O., Torbert, H. A., & Kloepper, J. W. (2010). Increased plant uptake of nitrogen from 15N-depleted fertilizer using plant growth-promoting rhizobacteria. *Applied Soil Ecology, 46*(1), 54–58. https://doi.org/10.1016/j.apsoil.2010.06.010

Agrios, G. N. (1995). *Fitopatología* (2nd ed.) (Limusa).

Aguilar-Bultet, L., & Falquet, L. (2015). Secuenciación y ensamblaje de novo de genomas bacterianos: Una alternativa para el estudio de nuevos patógenos. *Revista de Salud Animal, 37*(2), 125–132.

Ahemad, M., & Kibret, M. (2014). Mechanisms and applications of plant growth promoting rhizobacteria: Current perspective. *Journal of King Saud University - Science, 26*(1), 1–20. https://doi.org/10.1016/j.jksus.2013.05.001

Ahmed, E., & Holmström, S. J. M. (2014). Siderophores in environmental research: Roles and applications. *Microbial Biotechnology, 7*(3), 196–208. https://doi.org/10.1111/1751-7915.12117

Ahmed, S., Choudhury, A. R., Roy, S. K., Choi, J., Sayyed, R. Z., & Sa, T. (2021). Biomolecular painstaking utilization and assimilation of phosphorus under indigent stage in agricultural crops. In H. B. Singh, A. Vaishnav, & R. Z. Sayyed (Eds.), *Antioxidants in plant-microbe interaction* (pp. 565–588).

Akamatsu, H., Itoh, Y., Kodama, M., Otani, H., & Kohmoto, K. (1997). AAL-toxin-deficient mutants of *Alternaria alternata* tomato pathotype by restriction enzyme-mediated integration. *Phytopathology, 87*(9), 967–972. https://doi.org/10.1094/PHYTO.1997.87.9.967

Akimitsu, K., Tsuge, T., Kodama, M., Yamamoto, M., & Otani, H. (2014). Alternaria host-selective toxins: Determinant factors of plant disease. *Journal of General Plant Pathology, 80*(2), 109–122. https://doi.org/10.1007/s10327-013-0498-7

Akinrinlola, R. J., Yuen, G. Y., Drijber, R. A., & Adesemoye, A. O. (2018). Evaluation of *Bacillus* strains for plant growth promotion and predictability of efficacy by in vitro physiological traits. *International Journal of Microbiology.* https://doi.org/10.1155/2018/5686874, 2018.

Aktuganov, G. E., Galimzyanova, N. F., Melent'Ev, A. I., & Kuz'Mina, L. Y. (2007). Extracellular hydrolases of strain *Bacillus* sp. 739 and their involvement in the lysis of micromycete cell walls. *Microbiology, 76*(4), 413–420. https://doi.org/10.1134/S0026261707040054

Albareda, M., Rodríguez-Navarro, D. N., Camacho, M., & Temprano, F. J. (2008). Alternatives to peat as a carrier for rhizobia inoculants: Solid and liquid formulations. *Soil Biology and Biochemistry, 40*(11), 2771–2779. https://doi.org/10.1016/j.soilbio.2008.07.021

Alcantara, J., Acero, G., Alcantara, J., & Sánchez, M. (2019). Principales reguladores hormonales y sus interacciones en el crecimiento vegetal. *Nova, 17*(32), 109–129. http://www.scielo.org.co/pdf/nova/v17n32/1794-2470-nova-17-32-109.pdf.

Al-Khassawneh, N. M., Karam, N. S., & Shibli, R. A. (2006). Growth and flowering of black iris (*Iris nigricans* Dinsm.) following treatment with plant growth regulators. *Scientia Horticulturae, 107*(2), 187–193. https://doi.org/10.1016/j.scienta.2005.10.003

Allen, J., Davey, H. M., Broadhurst, D., Heald, J. K., Rowland, J. J., Oliver, S. G., & Kell, D. B. (2003). High-throughput classification of yeast mutants for functional genomics using metabolic footprinting. *Nature Biotechnology, 21*(6), 692–696. https://doi.org/10.1038/nbt823

Allouzi, M. M. A., Allouzi, S. M. A., Keng, Z. X., Supramaniam, C. V., Singh, A., & Chong, S. (2022). Liquid biofertilizers as a sustainable solution for agriculture. *Heliyon*, e12609.

Alori, E. T., & Babalola, O. O. (2018). Microbial inoculants for improving crop quality and human health in Africa. *Frontiers in Microbiology, 9*(September), 1–12. https://doi.org/10.3389/fmicb.2018.02213

Altschul, S. F., Gish, W., Miller, W., Myers, E. W., & Lipman, D. J. (1990). Basic local alignment search tool. *Journal of Molecular Biology, 215*(3), 403–410. https://doi.org/10.1016/S0022-2836(05)80360-2

Ambardar, S., Gupta, R., Trakroo, D., Lal, R., & Vakhlu, J. (2016). High throughput sequencing: An overview of sequencing chemistry. *Indian Journal of Microbiology, 56*(4), 394–404. https://doi.org/10.1007/s12088-016-0606-4

Ambrosini, A., de Souza, R., & Passaglia, L. M. P. (2016). Ecological role of bacterial inoculants and their potential impact on soil microbial diversity. *Plant and Soil, 400*(1–2), 193–207. https://doi.org/10.1007/s11104-015-2727-7

Anand, U., Vaishnav, A., Sharma, S. K., Sahu, J., Ahmad, S., Sunita, K., Suresh, S., Dey, A., Bontempi, E., Singh, A. K., Proćków, J., & Shukla, A. K. (2022). Current advances and research prospects for agricultural and industrial uses of microbial strains available in world collections. *Science of the Total Environment, 842*(January). https://doi.org/10.1016/j.scitotenv.2022.156641

Anaya, A., & Pedroza, H. (2008). Scaling-up, the art of chemical engineering: Pilot plants, the step between the egg and the hen. *Ciencia Ed. (IMIQ), 23*(1), 31–39.

Andrews, S. (2010). *FastQC: A quality control tool for high throughput sequence data.* https://www.bioinformatics.babraham.ac.uk/projects/fastqc/Jan2023.

Arahal, D. R. (2014). Whole-genome analyses: Average nucleotide identity. In *Methods in microbiology* (1st ed., Vol. 41). Elsevier Ltd. https://doi.org/10.1016/bs.mim.2014.07.002

Arjjumend, H., & Koutouki, K. (2021). Analysis of Indian and Canadian laws on biofertilizers. *Journal of Agricultural and Environmental Law, 16*(30), 7–23. https://doi.org/10.21029/jael.2021.30.7

Arkhipova, T. N., Prinsen, E., Veselov, S. U., Martinenko, E. V., Melentiev, A. I., & Kudoyarova, G. R. (2007). Cytokinin producing bacteria enhance plant growth in drying soil. *Plant and Soil, 292*(1–2), 305–315. https://doi.org/10.1007/s11104-007-9233-5

Aron, A. T., Gentry, E. C., McPhail, K. L., Nothias, L. F., Nothias-Esposito, M., Bouslimani, A., Petras, D., Gauglitz, J. M., Sikora, N., Vargas, F., van der Hooft, J. J. J., Ernst, M., Kang, K. Bin, Aceves, C. M., Caraballo-Rodríguez, A. M., Koester, I., Weldon, K. C., Bertrand, S.,

Roullier, C., et al. (2020). Reproducible molecular networking of untargeted mass spectrometry data using GNPS. *Nature Protocols, 15*(6), 1954–1991. https://doi.org/10.1038/s41596-020-0317-5

Arora, Naveen K, & Mishra, J. (2016). Prospecting the roles of metabolites and additives in future bioformulations for sustainable agriculture. *Applied Soil Ecology, 107*, 405–407. https://doi.org/10.1016/j.apsoil.2016.05.020

Arora, Naveen K., Khare, E., & Maheshwari, K. (2010). Plant growth promoting rhizobacteria: Constraints in bioformulation, commercialization, and future strategies. In D. K. Maheshwari (Ed.), *Plant growth and health promoting bacteria. Microbiology monographs* (Vol. 18, pp. 97–116). Springer. https://doi.org/10.1007/978-3-642-13612-2

Arrebola, E., Sivakumar, D., & Korsten, L. (2010). Effect of volatile compounds produced by *Bacillus* strains on postharvest decay in citrus. *Biological Control, 53*(1), 122–128. https://doi.org/10.1016/j.biocontrol.2009.11.010

Arroyo-Olarte, R. D., Bravo Rodríguez, R., & Morales-Ríos, E. (2021). Genome editing in bacteria: Crispr-cas and beyond. *Microorganisms, 9*(4). https://doi.org/10.3390/microorganisms9040844

Asil, M. H., Roein, Z., & Abbasi, J. (2011). Response of tuberose (*Polianthes tuberose* L.) to gibberellic acid and benzyladenine. *Horticulture Environment and Biotechnology, 52*(1), 46–51. https://doi.org/10.1007/s13580-011-0073-0

Atkins, S. D., Clark, I. M., Pande, S., Hirsch, P. R., & Kerry, B. R. (2005). The use of real-time PCR and species-specific primers for the identification and monitoring of *Paecilomyces lilacinus*. *FEMS Microbiology Ecology, 51*(2), 257–264. https://doi.org/10.1016/j.femsec.2004.09.002

Athukorala, S. N. P., & Fernando, W. G. D. (2009). Identification of antifungal antibiotics of bacillus species isolated from different microhabitats using polymerase chain reaction and maldi-TOF. *Mass Spectrometry, 1032*, 1021–1032. https://doi.org/10.1139/W09-067

Auch, A. F., Klenk, H. P., & Göker, M. (2010). Standard operating procedure for calculating genome-to-genome distances based on high-scoring segment pairs. *Standards in Genomic Sciences, 2*(1), 142–148. https://doi.org/10.4056/sigs.541628

Ayilara, M. S., Adeleke, B. S., & Babalola, O. O. (2022). Bioprospecting and challenges of plant microbiome research for sustainable agriculture, a review on soybean endophytic bacteria. *Microbial Ecology, 85*(3), 1113–1135.

Aziz, R. K., Bartels, D., Best, A., DeJongh, M., Disz, T., Edwards, R. A., Formsma, K., Gerdes, S., Glass, E. M., Kubal, M., Meyer, F., Olsen, G. J., Olson, R., Osterman, A. L., Overbeek, R. A., McNeil, L. K., Paarmann, D., Paczian, T., Parrello, B., et al. (2008). The RAST Server: Rapid annotations using subsystems technology. *BMC Genomics, 9*, 1–15. https://doi.org/10.1186/1471-2164-9-75

Bacilio, M., Moreno, M., Lopez-aguilar, D. R., & Bashan, Y. (2017). Scaling from the growth chamber to the greenhouse to the field: Demonstration of diminishing effects of mitigation of salinity in peppers inoculated with plant growth-promoting bacterium and humic acids. *Applied Soil Ecology, 119*(June), 327–338. https://doi.org/10.1016/j.apsoil.2017.07.002

Baetz, U., & Martinoia, E. (2014). Root exudates: The hidden part of plant defense. *Trends in Plant Science, 19*(2), 90–98. https://doi.org/10.1016/j.tplants.2013.11.006

Bahassi, E. M., & Stambrook, P. J. (2014). Next-generation sequencing technologies: Breaking the sound barrier of human genetics. *Mutagenesis, 29*(5), 303–310. https://doi.org/10.1093/mutage/geu031

Bakker, P. A. H. M., Berendsen, R. L., Doornbos, R. F., Wintermans, P. C. A., & Pieterse, C. M. J. (2013). The rhizosphere revisited: Root microbiomics. *Frontiers in Plant Science, 4*(May), 1–7. https://doi.org/10.3389/fpls.2013.00165

Bakri, Y., Akeed, Y., & Thonart, P. (2012). Comparison between continuous and batch processing to produce xylanase by *Penicillium canescens* 10-10c. *Brazilian Journal of Chemical Engineering, 29*(3), 441–447. https://doi.org/10.1590/S0104-66322012000300001

Banerjee, A., Bareh, D. A., & Joshi, S. R. (2017). Native microorganisms as potent bioinoculants for plant growth promotion in shifting agriculture (Jhum) systems. *Journal of Soil Science and Plant Nutrition, 17*(1), 127–140. https://doi.org/10.4067/S0718-95162017005000010

Bang-Andreasen, T., Anwar, M. Z., Lanzén, A., Kjøller, R., Rønn, R., Ekelund, F., & Jacobsen, C. S. (2019). Total RNA sequencing reveals multilevel microbial community changes and functional responses to wood ash application in agricultural and forest soil. *FEMS Microbiology Ecology, 96*(3), 1–13. https://doi.org/10.1093/femsec/fiaa016

Barahona, E., Navazo, A., Martínez-Granero, F., Zea-Bonilla, T., Pérez-Jiménez, R. M., Martín, M., & Rivilla, R. (2011). Pseudomonas fluorescens F113 mutant with enhanced competitive colonization ability and improved biocontrol activity against fungal root pathogens. *Applied and Environmental Microbiology, 77*(15), 5412–5419. https://doi.org/10.1128/.AEM.00320-11.

Bardas, G. A., Lagopodi, A. L., Kadoglidou, K., & Tzavella-klonari, K. (2009). Biological control of three *Colletotrichum lindemuthianum* races using *Pseudomonas chlororaphis* PCL1391 and *Pseudomonas fluorescens* WCS365. *Biological Control, 49*(2), 139–145. https://doi.org/10.1016/j.biocontrol.2009.01.012

Barea, J. M. (2015). Future challenges and perspectives for applying microbial biotechnology in sustainable agriculture based on a better understanding of plant-microbiome interactions. *Journal of Soil Science and Plant Nutrition, 15*(2), 261–282. https://doi.org/10.4067/s0718-95162015005000021

Bari, R., & Jones, J. D. G. (2009). Role of plant hormones in plant defence responses. *Plant Molecular Biology, 69*(4), 473–488. https://doi.org/10.1007/s11103-008-9435-0

Bar-On, Y. M., Phillips, R., & Milo, R. (2018). The biomass distribution on Earth. *Proceedings of the National Academy of Sciences of the United States of America, 115*(25), 6506–6511. https://doi.org/10.1073/pnas.1711842115

Barquero, M., Pastor-Buies, R., Urbano, B., & González-Andrés, F. (2019). Challenges, regulations and future actions in biofertilizers in the European agriculture: From the lab to the field. In D. Zúñiga-Dávila, F. González-Andrés, & E. Ormeño-Orrillo (Eds.), *Microbial probiotics for agricultural systems* (pp. 83–107). Springer. https://doi.org/10.1007/978-3-030-17597-9_6

Barraclough, D. (1995). 15N isotope dilution techniques to study soil nitrogen transformations and plant uptake. *Fertilizer Research, 42*, 185–192.

Bashan, Y., Hernandez, J. P., Leyva, L. A., & Bacilio, M. (2002). Alginate microbeads as inoculant carriers for plant growth-promoting bacteria. *Biology and Fertility of Soils, 35*(5), 359–368. https://doi.org/10.1007/s00374-002-0481-5

Bashan, Y., Salazar, B. G., Moreno, M., Lopez, B. R., & Linderman, R. G. (2012). Restoration of eroded soil in the Sonoran desert with native leguminous trees using plant growth-promoting microorganisms and limited amounts of compost and water. *Journal of Environmental Management, 102*, 26–36. https://doi.org/10.1016/j.jenvman.2011.12.032

de-Bashan, L. E., Hernandez, J. P., & Bashan, Y. (2012). The potential contribution of plant growth-promoting bacteria to reduce environmental degradation - a comprehensive evaluation. *Applied Soil Ecology, 61*, 171–189. https://doi.org/10.1016/j.apsoil.2011.09.003

Bashan, Y., de-Bashan, L. E., & Prabhu, S. R. (2016). Superior polymeric formulations and emerging innovative products of bacterial inoculants for sustainable agriculture and the environment. In H. B. Singh, B. K. Sarma, & C. Keswani (Eds.), *Agriculturally important microorganisms: Commercialization and regulatory requirements in asia* (pp. 1–305). https://doi.org/10.1007/978-981-10-2576-1

Bashan, Y., de-Bashan, L. E., Prabhu, S. R., & Hernandez, J. P. (2014). Advances in plant growth-promoting bacterial inoculant technology: Formulations and practical perspectives (1998–2013). *Plant and Soil, 378*, 1–33. https://doi.org/10.1007/s11104-013-1956-x

Basse, C. W., Stumpferl, S., & Kahmann, R. (2000). Characterization of a ustilago maydis gene specifically induced during the biotrophic phase: Evidence for negative as well as positive regulation. *Molecular and Cellular Biology, 20*(1), 329–339. https://doi.org/10.1128/mcb.20.1.329-339.2000

Baudoin, E., Couillerot, O., Spaepen, S., Moënne-Loccoz, Y., & Nazaret, S. (2010). Applicability of the 16S-23S rDNA internal spacer for PCR detection of the phytostimulatory PGPR inoculant *Azospirillum lipoferum* CRT1 in field soil. *Journal of Applied Microbiology, 108*(1), 25–38. https://doi.org/10.1111/j.1365-2672.2009.04393.x

Becker, P., Bosschaerts, M., Chaerle, P., Daniel, H.-M., Hellemans, A., Olbrechts, A., Rigouts, L., Wilmotte, A., & Hendrickx, M. (2019). Public microbial resource centers: Key hubs for findable, accessible, interoperable, and reusable (FAIR) microorganisms and genetic materials. *Applied and Environmental Microbiology, 85*(21), e1444–e1459.

Beltrán-Pineda, M. E. (2014). La solubilización de fosfatos como estrategia microbiana para promover el crecimiento vegetal phosphate solubilization as a microbial strategy for promoting plant growth. *Corpoica Ciencia Tecnologia Agropecuaria, 15*(1), 101–113.

Benavides, L., Sproviero, C., Larrandart, A., García Alba, M., Ibarra, C., Bozzo, M., Solís, A., & Malter Terrada, M. (2019). Evaluation of nitrogen fertilization in sweet corn under sustainable management practices. *XLII annual meeting of the Argentine association of nuclear technology (AATN 2017)*.

Beneduzi, A., Ambrosini, A., & Passaglia, L. M. P. (2012). Plant growth-promoting rhizobacteria (PGPR): Their potential as antagonists and biocontrol agents. *Genetics and Molecular Biology, 35*(4), 1044–1051. https://doi.org/10.1590/S1415-47572012000600020

Bennett, S. (2004). Solexa Ltd. *Pharmacogenomics, 5*(4), 433–438. https://doi.org/10.1517/14622416.5.4.433

Berg, G., Kusstatscher, P., Abdelfattah, A., Cernava, T., & Smalla, K. (2021). Microbiome modulation—toward a better understanding of plant microbiome response to microbial inoculants. *Frontiers in Microbiology, 12*(April), 1–12. https://doi.org/10.3389/fmicb.2021.650610

Berlanga-Clavero, M. V., Molina-Santiago, C., Caraballo-Rodríguez, A. M., Petras, D., Díaz-Martínez, L., Pérez-García, A., de Vicente, A., Carrión, V. J., Dorrestein, P. C., & Romero, D. (2022). *Bacillus subtilis* biofilm matrix components target seed oil bodies to promote growth and anti-fungal resistance in melon. *Nature Microbiology, 7*(7), 1001–1015. https://doi.org/10.1038/s41564-022-01134-8

Berninger, T., González López, Ó., Bejarano, A., Preininger, C., & Sessitsch, A. (2018). Maintenance and assessment of cell viability in formulation of non-sporulating bacterial inoculants. *Microbial Biotechnology, 11*(2), 277–301. https://doi.org/10.1111/1751-7915.12880

Berovič, M. (2011). Sterilization in biotechnology. In M.-Y. Murray (Ed.), *Comprehensive biotechnology* (Vol. 2, pp. 135–150). Elsevier B.V. https://doi.org/10.1016/B978-0-08-088504-9.00093-3. Second.

Bertels, F., Silander, O. K., Pachkov, M., Rainey, P. B., & Van Nimwegen, E. (2014). Automated reconstruction of whole-genome phylogenies from short-sequence reads. *Molecular Biology and Evolution, 31*(5), 1077–1088. https://doi.org/10.1093/molbev/msu088

Bhalla, R., Dalal, M., Panguluri, S. K., Jagadish, B., Mandaokar, A. D., Singh, A. K., & Kumar, P. A. (2005). Isolation, characterization and expression of a novel vegetative insecticidal protein gene of *Bacillus thuringiensis*. *FEMS Microbiology Letters, 243*(2), 467–472. https://doi.org/10.1016/j.femsle.2005.01.011

Bharathi, R., Vivekananthan, R., Harish, S., Ramanathan, A., & Samiyappan, R. (2004). Rhizobacteria-based bio-formulations for the management of fruit rot infection in chillies. *Crop Protection, 23*, 835–843. https://doi.org/10.1016/j.cropro.2004.01.007

Bharti, N., & Suryavanshi, M. (2021). Quality control and regulations of biofertilizers: Current scenario and future prospects. In *Biofertilizers: Volume 1: Advances in bio-inoculants* (pp. 133–141). Woodhead Publishing. https://doi.org/10.1016/B978-0-12-821667-5.00018-X

Bhatia, S., & Bera, T. (2015). Classical and nonclassical techniques for secondary metabolite production in plant cell culture. In *Modern applications of plant biotechnology in pharmaceutical sciences*. Elsevier Inc. https://doi.org/10.1016/B978-0-12-802221-4.00007-8

Bhattacharyya, P. N., & Jha, D. K. (2012). Plant growth-promoting rhizobacteria (PGPR): Emergence in agriculture. *World Journal of Microbiology and Biotechnology, 28*(4), 1327–1350. https://doi.org/10.1007/s11274-011-0979-9

Bhattarai, B. (2015). Variation of soil microbial population in different soil horizons. *Journal of Microbiology and Experimentation, 2*(2), 75–78. https://doi.org/10.15406/jmen.2015.02.00044

Bhuiyan, S. A., Garlick, K., Anderson, J. M., Wickramasinghe, P., & Stirling, G. R. (2018). Biological control of root-knot nematode on sugarcane in soil naturally or artificially infested with *Pasteuria penetrans*. *Australasian Plant Pathology, 47*(1), 45–52. https://doi.org/10.1007/s13313-017-0530-z

Bienert, M. D., Baijot, A., & Boutry, M. (2014). *ABCG transporters and their role in the biotic stress response* (pp. 137–162). https://doi.org/10.1007/978-3-319-06511-3_8

Bigeard, J., & Hirt, H. (2018). Nuclear signaling of plant MAPKs. *Frontiers in Plant Science, 9*(April), 1–18. https://doi.org/10.3389/fpls.2018.00469

Bigeard, J., Colcombet, J., & Hirt, H. (2015). Signaling mechanisms in pattern-triggered immunity (PTI). *Molecular Plant, 8*(4), 521–539. https://doi.org/10.1016/j.molp.2014.12.022

Bilgin, D. D., Zavala, J. A., Zhu, J., Clough, S. J., Ort, D. R., & Delucia, E. H. (2010). Biotic stress globally downregulates photosynthesis genes. *Plant, Cell and Environment, 33*(10), 1597–1613. https://doi.org/10.1111/j.1365-3040.2010.02167.x

Bissonette, N., Lalande, R., & Bordeleau, L. M. (1986). Large-scale production of *Rhizobium meliloti* on whey. *Applied and Environmental Microbiology, 52*(4), 838–841. https://doi.org/10.1128/aem.52.4.838-841.1986

Blin, K., Shaw, S., Kloosterman, A. M., Charlop-Powers, Z., Van Wezel, G. P., Medema, M. H., & Weber, T. (2021). AntiSMASH 6.0: improving cluster detection and comparison capabilities. *Nucleic Acids Research, 49*. https://doi.org/10.1093/nar/gkab335

Bohn, G. W., & Tucker, C. M. (1939). Immunity to fusarium wilt in the tomato. *Science, 89*(2322), 603–604. https://doi.org/10.1126/science.89.2322.603

Bolger, A. M., Lohse, M., & Usadel, B. (2014). Trimmomatic: A flexible trimmer for illumina sequence data. *Bioinformatics, 30*(15), 2114–2120. https://doi.org/10.1093/bioinformatics/btu170

Borriss, R. (2011). Bacteria in agrobiology: Plant growth responses. In *Bacteria in agrobiology: Plant growth responses*. https://doi.org/10.1007/978-3-642-20332-9

Brahmaprakash, G. P., & Sahu, P. K. (2012). Biofertilizers for sustainability. *Journal of the Indian Institute of Science, 92*(1), 37–62.

Brahmaprakash, G. P., Sahu, P. K., Lavanya, G., Gupta, A., Nair, S. S., & Gangaraddi, V. (2020). Role of additives in improving efficiency of bioformulation for plant growth and development. In *Frontiers in soil and environmental microbiology* (pp. 1–10). CRC Press.

Brar, S. K., Verma, M., Tyagi, R. D., & Valéro, J. R. (2006). Recent advances in downstream processing and formulations of *Bacillus thuringiensis* based biopesticides. *Process Biochemistry, 41*(2), 323–342. https://doi.org/10.1016/j.procbio.2005.07.015

Briand, M., Bouzid, M., Hunault, G., Legeay, M., Fischer-Le Saux, M., & Barret, M. (2021). A rapid and simple method for assessing and representing genome sequence relatedness. *Peer Community Journal, 1*, 1–14. https://doi.org/10.24072/pcjournal.37

Briones-Moreno, A., Hernández-García, J., Vargas-Chávez, C., Romero-Campero, F. J., Romero, J. M., Valverde, F., & Blázquez, M. A. (2017). Evolutionary analysis of DELLA-associated transcriptional networks. *Frontiers in Plant Science, 8*(April), 1–11. https://doi.org/10.3389/fpls.2017.00626

Buermans, H. P. J., & den Dunnen, J. T. (2014). Next generation sequencing technology: Advances and applications. *Biochimica et Biophysica Acta - Molecular Basis of Disease, 1842*(10), 1932–1941. https://doi.org/10.1016/j.bbadis.2014.06.015

Bussas, V., Sharma, A., & Shouche, Y. (2017). IP and the budapest treaty-depositing biological material for patent purposes. In *Microbial resources: from functional existence in nature to applications (issue august 2019)*. Elsevier Inc. https://doi.org/10.1016/B978-0-12-804765-1.00014-X

Bustin, S., & Huggett, J. (2017). qPCR primer design revisited. *Biomolecular Detection and Quantification, 14*, 19–28. https://doi.org/10.1016/j.bdq.2017.11.001

Cai, Q., Qiao, L., Wang, M., He, B., Lin, F., Palmquist, J., & Jin, H. (2018). Pathogen to silence virulence genes. *Science, 360*(June), 1126–1129.

Cai, Q., He, B., Weiberg, A., Buck, A. H., & Jin, H. (2019). Small RNAs and extracellular vesicles: New mechanisms of cross-species communication and innovative tools for disease control. *PLoS Pathogens, 15*(12), 1–13. https://doi.org/10.1371/journal.ppat.1008090

Calvo, P., Nelson, L., & Kloepper, J. W. (2014). Agricultural uses of plant biostimulants. *Plant and Soil, 383*(1–2), 3–41. https://doi.org/10.1007/s11104-014-2131-8

Camilios-Neto, D., Bonato, P., Wassem, R., Tadra-Sfeir, M. Z., Brusamarello-Santos, L. C. C., Valdameri, G., Donatti, L., Faoro, H., Weiss, V. A., Chubatsu, L. S., Pedrosa, F. O., & Souza, E. M. (2014). Dual RNA-seq transcriptional analysis of wheat roots colonized by *Azospirillum brasilense* reveals up-regulation of nutrient acquisition and cell cycle genes. *BMC Genomics, 15*(1), 1–13. https://doi.org/10.1186/1471-2164-15-378

Campos-Soriano, L., García-Martínez, J., & Segundo, B. S. (2012). The arbuscular mycorrhizal symbiosis promotes the systemic induction of regulatory defence-related genes in rice leaves and confers resistance to pathogen infection. *Molecular Plant Pathology, 13*(6), 579–592. https://doi.org/10.1111/j.1364-3703.2011.00773.x

Cao, Y., Pi, H., Chandrangsu, P., Li, Y., Wang, Y., Zhou, H., Xiong, H., Helmann, J. D., & Cai, Y. (2018). Antagonism of two plant-growth promoting *Bacillus velezensis* isolates against ralstonia solanacearum and *Fusarium oxysporum. Scientific Reports, 8*(1), 1–14. https://doi.org/10.1038/s41598-018-22782-z

Cardozo, P., Di Palma, A., Martin, S., Cerliani, C., Esposito, G., Reinoso, H., & Travaglia, C. (2022). Improvement of maize yield by foliar application of *Azospirillum brasilense* Az39. *Journal of Plant Growth Regulation, 41*(3), 1032–1040. https://doi.org/10.1007/s00344-021-10356-9

Castro-moretti, F. R., Gentzel, I. N., Mackey, D., & Alonso, A. P. (2020). Metabolomics as an emerging tool for the study of plant–pathogen interactions. *Metabolites, 10*(2), 1–23. https://doi.org/10.3390/metabo10020052

Catanzariti, A. M., Do, H. T. T., Bru, P., de Sain, M., Thatcher, L. F., Rep, M., & Jones, D. A. (2017). The tomato I gene for *Fusarium* wilt resistance encodes an atypical leucine-rich repeat receptor-like protein whose function is nevertheless dependent on SOBIR1 and SERK3/BAK1. *Plant Journal, 89*(6), 1195–1209. https://doi.org/10.1111/tpj.13458

Cawoy, H., Debois, D., Franzil, L., Pauw, E. De, Thonart, P., & Ongena, M. (2014). *Lipopeptides as main ingredients for inhibition of fungal phytopathogens by* Bacillus subtilis*/amyloliquefaciens* (Vol. 12). https://doi.org/10.1111/1751-7915.12238

Chaiharn, M., Chunhaleuchanon, S., & Lumyong, S. (2009). Screening siderophore producing bacteria as potential biological control agent for fungal rice pathogens in Thailand. *World Journal of Microbiology and Biotechnology, 25*(11), 1919–1928. https://doi.org/10.1007/s11274-009-0090-7

Chang, W. T., Hsieh, C. H., Hsieh, H. S., & Chen, C. (2009). Conversion of crude chitosan to an anti-fungal protease by *Bacillus cereus. World Journal of Microbiology and Biotechnology, 25*(3), 375–382. https://doi.org/10.1007/s11274-008-9901-5

Chang, Y. S., & Sung, F. H. (2000). Effects of gibberellic acid and dormancy-breaking chemicals on flower development of *Rhododendron pulchrum* Sweet and R. scabrum Don. *Scientia Horticulturae, 83*(3–4), 331–337. https://doi.org/10.1016/S0304-4238(99)00111-9

Chaparro-Encinas, L. A., Parra-Cota, F. I., Cruz-Mendívil, A., Santoyo, G., Peña-Cabriales, J. J., Castro-Espinoza, L., & de los Santos-Villalobos, S. (2022). Transcriptional regulation of cell growth and reprogramming of systemic response in wheat (*Triticum turgidum* subsp. durum) seedlings by Bacillus paralicheniformis TRQ65. *Planta, 255*(3), 1–15. https://doi.org/10.1007/s00425-022-03837-y

Chaparro-Encinas, L. A., Rojas-Padilla, J., Valenzuela-Aragón, B., Santoyo-Pizano, G., Orozco-Mosqueda, M. C., Parra-Cota, F. I., & de los Santos-Villalobos, S. (2020). Respuesta molecular de las plantas a la inoculación de bacterias promotoras del crecimiento vegetal del género Bacillus. In M. C. Orozco-Mosqueda, & G. Santoyo-Pizano (Eds.), *Bacterias promotoras del crecimiento vegetal. Aspectos básicos y aplicaciones para una agricultura sustentable* (pp. 83–109). Editorial Fontamara, S. A. de C. V, ISBN 978-607-736-659-1.

Chaudhari, Ami, & Patel, J. (2021). Current status and future perspective on enzyme involving in biocontrol of plant pathogen. *Journal for Research in Applied Sciences and Biotechnology, 8*(4), 49–55. https://doi.org/10.55544/jrasb.1.1.3

Chaudhary, T., Dixit, M., Gera, R., Shukla, A. K., Prakash, A., Gupta, G., & Shukla, P. (2020). Techniques for improving formulations of bioinoculants. *3 Biotech, 10*, 199. https://doi.org/10.1007/s13205-020-02182-9

Chávez-Luzanía, R. A., Montoya-Martínez, A. C., Cota, F. I. P., & de los Santos-Villalobos, S. (2022). Pangenomes - identified singletons for designing specific primers to identify bacterial strains in a plant growth - promoting consortium. *Molecular Biology Reports, 49*, 10489–10498.

Chen, Y., Wang, J., Yang, N., Wen, Z., Sun, X., Chai, Y., & Ma, Z. (2018). Wheat microbiome bacteria can reduce virulence of a plant pathogenic fungus by altering histone acetylation. *Nature Communications, 9*(1), 1–14. https://doi.org/10.1038/s41467-018-05683-7

Chen, T., Nomura, K., Wang, X., Sohrabi, R., Xu, J., Yao, L., Paasch, B. C., Ma, L., Kremer, J., Cheng, Y., Zhang, L., Wang, N., Wang, E., Xin, X. F., & He, S. Y. (2020). A plant genetic network for preventing dysbiosis in the phyllosphere. *Nature, 580*(7805), 653–657. https://doi.org/10.1038/s41586-020-2185-0

Chiapello, M., Perotto, S., & Balestrini, R. (2015). Symbiotic proteomics — state of the art in plant–mycorrhizal fungi interactions. *Recent Advances in Proteomics Research.* . https://doi.org/10.5772/61331

Chin, C. S., Peluso, P., Sedlazeck, F. J., Nattestad, M., Concepcion, G. T., Clum, A., Dunn, C., O'Malley, R., Figueroa-Balderas, R., Morales-Cruz, A., Cramer, G. R., Delledonne, M.,

Luo, C., Ecker, J. R., Cantu, D., Rank, D. R., & Schatz, M. C. (2016). Phased diploid genome assembly with single-molecule real-time sequencing. *Nature Methods, 13*(12), 1050–1054. https://doi.org/10.1038/nmeth.4035

Chin-a-woeng, T. F. C., Bloemberg, G. V., Bij, A. J. Van Der, Drift, K. M. G. M. Van Der, Schripsema, J., Kroon, B., Scheffer, R. J., Keel, C., Bakker, P. A. H. M., Tichy, H., Bruijn, F. J. De, Thomas-oates, J. E., & Lugtenberg, B. J. J. (1998). Biocontrol by Phenazine-1-carboxamide-producing pseudomonas chlororaphis PCL1391 of tomato root rot caused by fusarium oxysporum F . Sp . radicis-lycopersici. *Molecular Plant Microbe Interactions, 11*(11), 1069–1077.

Chisholm, S. T., Coaker, G., Day, B., & Staskawicz, B. J. (2006). Host-microbe interactions: Shaping the evolution of the plant immune response. *Cell, 124*(4), 803–814. https://doi.org/10.1016/j.cell.2006.02.008

Choudhary, D. K., & Johri, B. N. (2009). Interactions of *Bacillus* spp. and plants - with special reference to induced systemic resistance (ISR). *Microbiological Research, 164*(5), 493–513. https://doi.org/10.1016/j.micres.2008.08.007

Chowdhury, S. P., Uhl, J., Grosch, R., Alquéres, S., Pittroff, S., Dietel, K., Schmitt-Kopplin, P., Borriss, R., & Hartmann, A. (2015). Cyclic lipopeptides of *Bacillus amyloliquefaciens* subsp. plantarum colonizing the lettuce rhizosphere enhance plant defense responses toward the bottom rot pathogen *Rhizoctonia solani. Molecular Plant-Microbe Interactions, 28*(9), 984–995. https://doi.org/10.1094/MPMI-03-15-0066-R

Chowdhury, S., Basu, A., & Kundu, S. (2017). Biotrophy-necrotrophy switch in pathogen evoke differential response in resistant and susceptible sesame involving multiple signaling pathways at different phases. *Scientific Reports, 7*(1), 1–17. https://doi.org/10.1038/s41598-017-17248-7

Christian, M., Qi, Y., Zhang, Y., & Voytas, D. F. (2013). Targeted mutagenesis of *Arabidopsis thaliana* using engineered TAL effector nucleases. *G3: Genes, Genomes, Genetics, 3*(9), 1697–1705. https://doi.org/10.1534/g3.113.007104

Chun, J., & Rainey, F. A. (2014). Integrating genomics into the taxonomy and systematics of the bacteria and archaea. *International Journal of Systematic and Evolutionary Microbiology, 64*, 316–324. https://doi.org/10.1099/ijs.0.054171-0

Chun, J., Oren, A., Ventosa, A., Christensen, H., Arahal, D. R., da Costa, M. S., Rooney, A. P., Yi, H., Xu, X. W., De Meyer, S., & Trujillo, M. E. (2018). Proposed minimal standards for the use of genome data for the taxonomy of prokaryotes. *International Journal of Systematic and Evolutionary Microbiology, 68*(1), 461–466. https://doi.org/10.1099/ijsem.0.002516

Cohen, A. C., Travaglia, C. N., Bottini, R., & Piccoli, P. N. (2009). Participation of abscisic acid and gibberellins produced by endophytic azospirillum in the alleviation of drought effects in maize. *Botany, 87*(5), 455–462. https://doi.org/10.1139/B09-023

Commare, R. R., Nandakumar, R., Kandan, A., Suresh, S., Bharathi, M., Raguchandera, T., & Samiyappana, R. (2002). Pseudomonas fluorescens based bio-formulation for the management of sheath blight disease and leaffolder insect in rice. *Crop Protection, 21*, 671–677.

Compant, S., Samad, A., Faist, H., & Sessitsch, A. (2019). A review on the plant microbiome: Ecology, functions, and emerging trends in microbial application. *Journal of Advanced Research, 19*, 29–37. https://doi.org/10.1016/j.jare.2019.03.004

Concu, R., & Cordeiro, M. N. D. S. (2019). Alignment-free method to predict enzyme classes and subclasses. *International Journal of Molecular Sciences, 20*(21). https://doi.org/10.3390/ijms20215389

Conga, P. T., Dunga, T. D., Hiena, T. M., Hienb, N. T., Choudhuryc, A. T. M. A., Kecskés, M. L., & Kennedy, I. R. (2009). Inoculant plant growth-promoting microorganisms enhance utilisation of urea-N and grain yield of paddy rice in southern Vietnam. *European Journal of Soil Biology, 45*, 52–61. https://doi.org/10.1016/j.ejsobi.2008.06.006

Córdova-Albores, L. C., Zelaya-Molina, L. X., Ávila-Alistac, N., Valenzuela-Ruíz, V., Cortés-Martínez, N. E., Parra-Cota, F. I., Burgos-Canul, Y. Y., Chávez-Díaz, I. F., Fajardo-Franco, M. L., & De los Santos-Villalobos, S. (2020). Omics sciences potential on bio-prospecting of biological control microbial agents: The case of the Mexican agro-biotechnology. *Mexican Journal of Phytopathology, 39*(1), 147–184. https://doi.org/10.18781/r.mex.fit.2009-3

Cornforth, D. M., Popat, R., McNally, L., Gurney, J., Scott-Phillips, T. C., Ivens, A., Diggle, S. P., & Brown, S. P. (2014). Combinatorial quorum sensing allows bacteria to resolve their social and physical environment. *Proceedings of the National Academy of Sciences of the United States of America, 111*(11), 4280–4284. https://doi.org/10.1073/pnas.1319175111

Cortés-Patiño, S., & Bonilla, R. R. (2015). Polymers selection for a liquid inoculant of *Azospirillum brasilense* based on the arrhenius thermodynamic model. *African Journal of Biotechnology, 14*(33), 2547–2553. https://doi.org/10.5897/AJB2015.14777

Cossus, L., Roux-Dalvai, F., Kelly, I., Nguyen, T. T. A., Antoun, H., Droit, A., & Tweddell, R. J. (2021). Interactions with plant pathogens influence lipopeptides production and antimicrobial activity of *Bacillus subtilis* strain PTB185. *Biological Control, 154*(March), 104497. https://doi.org/10.1016/j.biocontrol.2020.104497

Couillerot, O., Bouffaud, M.-L., Baudoin, E., Muller, D., Caballero-Mellado, J., & Moënne-Loccoz, Y. (2010). Development of a real-time PCR method to quantify the PGPR strain *Azospirillum lipoferum* CRT1 on maize seedlings. *Soil Biology and Biochemistry, 42*(12), 2298–2305. https://doi.org/10.1016/j.soilbio.2010.09.003

Cubasch, U., & Wuebbles, D. (2013). Introduction: Climate change. In T. F. Stocker, D. Qin, G.-K. Plattner, M. Tignor, S. K. Allen, J. Boschung, A. Nauels, Y. Xia, V. Bex, & P. M. Midgley (Eds.), *Climate change 2013: The physical science basis. Contribution of working group I to the fifth assessment report of the intergovernmental panel on climate change* (pp. 119–158). Cambridge University Press. https://doi.org/10.1016/B978-0-12-809665-9.15009-8

Dangl', J. L., Dietrich, R. A., & Richberg', M. H. (1996). Death don't have No mercy: Cell death programs in plant-microbe lnteractions. In *The plant cell* (Vol. 8)American Society of Plant Physiologists. https://academic.oup.com/plcell/article/8/10/1793/5985248.

Das, S., Dash, H. R., Mangwani, N., Chakraborty, J., & Kumari, S. (2014). Understanding molecular identification and polyphasic taxonomic approaches for genetic relatedness and phylogenetic relationships of microorganisms. *Journal of Microbiological Methods, 103*, 80–100. https://doi.org/10.1016/j.mimet.2014.05.013

Dastogeer, K. M., Tumpa, F. H., Sultana, A., Akter, M. A., & Chakraborty, A. (2020). Plant microbiome—an account of the factors that shape community composition and diversity. *Current Plant Biology, 23*, 100161.

Dauner, M., Storni, T., & Sauer, U. (2001). *Bacillus subtilis* metabolism and energetics in carbon-limited and excess-carbon chemostat culture. *Journal of Bacteriology, 183*(24), 7308–7317. https://doi.org/10.1128/JB.183.24.7308-7317.2001

De Coninck, B., Timmermans, P., Vos, C., Cammue, B. P. A., & Kazan, K. (2015). What lies beneath: Belowground defense strategies in plants. *Trends in Plant Science, 20*(2), 91–101. https://doi.org/10.1016/j.tplants.2014.09.007

De Souza, K. P., Setubal, J. C., André Carlos, A. C. P., Oliveira, G., Chateau, A., & Alves, R. (2019). Machine learning meets genome assembly. *Briefings in Bioinformatics, 20*(6), 2116–2129. https://doi.org/10.1093/bib/bby072

De Vero, L., Boniotti, M. B., Budroni, M., Buzzini, P., Cassanelli, S., Comunian, R., Gullo, M., Logrieco, A. F., Mannazzu, I., Musumeci, R., Perugini, I., Perrone, G., Pulvirenti, A., Romano, P., Turchetti, B., & Varese, G. C. (2019). Preservation, characterization and exploitation of microbial biodiversity: The perspective of the Italian network of culture collections. *Microorganisms, 7*(12). https://doi.org/10.3390/microorganisms7120685

De Vleesschauwer, D., & Höfte, M. (2009). Chapter 6 rhizobacteria-induced systemic resistance. *Advances in Botanical Research, 51*(C), 223–281. https://doi.org/10.1016/S0065-2296(09)51006-3

De Vleesschauwer, D., Djavaheri, M., Bakker, P. A. H. M., & Höfte, M. (2008). Pseudomonas fluorescens WCS374r-induced systemic resistance in rice against *Magnaporthe oryzae* is based on pseudobactin-mediated priming for a salicylic acid-repressible multifaceted defense response. *Plant Physiology, 148*(4), 1996–2012. https://doi.org/10.1104/pp.108.127878

Deaker, R., Roughley, R. J., & Kennedy, I. R. (2004). Legume seed inoculation technology - a review. *Soil Biology and Biochemistry, 36*(8), 1275–1288. https://doi.org/10.1016/j.soilbio.2004.04.009

Debois, D., Jourdan, E., Smargiasso, N., Thonart, P., De Pauw, E., & Ongena, M. (2014). Spatiotemporal monitoring of the antibiome secreted by bacillus biofilms on plant roots using MALDI mass spectrometry imaging. *Analytical Chemistry, 86*(9), 4431–4438. https://doi.org/10.1021/ac500290s

Dehghaniana, S. Z., Abdollahi, M., Charehgani, H., & Niazi, A. (2019). Combined of salicylic acid and Pseudomonas fl uorescens CHA0 on the expression of PR1 gene and control of *Meloidogyne javanica* in tomato. *Biological Control*, 104134. https://doi.org/10.1016/j.biocontrol.2019.104134. July.

Deloger, M., El Karoui, M., & Petit, M. A. (2009). A genomic distance based on MUM indicates discontinuity between most bacterial species and genera. *Journal of Bacteriology, 91*(1), 91–99. https://doi.org/10.1128/JB.01202-08

Dempsey, D. A., & Klessig, D. F. (2012). SOS - too many signals for systemic acquired resistance? *Trends in Plant Science, 17*(9), 538–545. https://doi.org/10.1016/j.tplants.2012.05.011

Depuydt, S., Trenkamp, S., Fernie, A. R., Elftieh, S., Renou, J.-P., Vuylsteke, M., Holsters, M., & Vereecke, D. (2009). An integrated genomics approach to define niche establishment by rhodococcus fascians. *Plant Physiology, 149*(3), 1366–1386. https://doi.org/10.1104/pp.108.131805

Desaki, Y., Kohari, M., Shibuya, N., & Kaku, H. (2019). MAMP-triggered plant immunity mediated by the LysM-receptor kinase CERK1. *Journal of General Plant Pathology, 85*(1). https://doi.org/10.1007/s10327-018-0828-x

Desjardins, A. E., & Hohn, T. M. (1997). Mycotoxins in plant pathogenesis. *Molecular Plant-Microbe Interactions, 10*(2), 147–152. https://doi.org/10.1094/MPMI.1997.10.2.147

Desjardins, A. E., Proctor, R. H., Bai, G., McCormick, S. P., Shaner, G., Buechley, G., & Hohn, T. M. (1996). Reduced virulence of trichothecene-nonproducing mutants of *Gibberella zeae* in wheat field tests. *Molecular Plant-Microbe Interactions, 9*(9), 775–781. https://doi.org/10.1094/MPMI-9-0775

Desmeth, P. (2017). The nagoya protocol applied to microbial genetic resources. In *Microbial resources: From functional existence in nature to applications*. Elsevier Inc. https://doi.org/10.1016/B978-0-12-804765-1.00010-2

Di Benedetto, N. A., Corbo, M. R., Campaniello, D., Cataldi, M. P., Bevilacqua, A., Sinigaglia, M., & Flagella, Z. (2017). The role of plant growth promoting bacteria in improving nitrogen use efficiency for sustainable crop production: A focus on wheat. *AIMS Microbiology, 3*(3), 413–434. https://doi.org/10.3934/microbiol.2017.3.413

Di Benedetto, N. A., Campaniello, D., Bevilacqua, A., Cataldi, M. P., Sinigaglia, M., Flagella, Z., & Corbo, M. R. (2019). Isolation, screening, and characterization of plant-growth-promoting bacteria from durum wheat rhizosphere to improve N and P nutrient use efficiency. *Microorganisms, 7*(11). https://doi.org/10.3390/microorganisms7110541

Di Francesco, A., & Baraldi, E. (2021). How siderophore production can influence the biocontrol activity of *Aureobasidium pullulans* against *Monilinia laxa* on peaches. *Biological Control, 152*, 104456. https://doi.org/10.1016/j.biocontrol.2020.104456

Di Salvo, L. P., Cellucci, G. C., Carlino, M. E., & García de Salamone, I. E. (2018). Plant growth-promoting rhizobacteria inoculation and nitrogen fertilization increase maize (*Zea mays* L.) grain yield and modified rhizosphere microbial communities. *Applied Soil Ecology, 126*(December), 113–120. https://doi.org/10.1016/j.apsoil.2018.02.010

Díaz-Puentes, L. (2012). Systemic acquired resistance induced by salicylic acid resistência sistêmica adquirida. *Biotecnología En El Sector Agropecuario Y Agroindustrial, 10*(2), 257–267.

Díaz-Rodríguez, María, Alondra, Parra-Cota, F. I., Santoyo, G., & de los Santos-Villalobos, S. (2019). Chlorothalonil tolerance of indole producing bacteria associated to wheat (*Triticum turgidum* L.) rhizosphere in the Yaqui Valley, Mexico. *Ecotoxicology, 28*(5), 569–577. https://doi.org/10.1007/s10646-019-02053-x

Díaz-Rodríguez, A. M., Salcedo Gastelum, L. A., Félix Pablos, C. M., Parra-Cota, F. I., Santoyo, G., Puente, M. L., Bhattacharya, D., Mukherjee, J., & de los Santos-Villalobos, S. (2021). The current and future role of microbial culture collections in food security worldwide. *Frontiers in Sustainable Food Systems, 4*. https://doi.org/10.3389/fsufs.2020.614739

Dicenzo, G. C., Zamani, M., Checcucci, A., Fondi, M., Griffitts, J. S., Finan, T. M., & Mengoni, A. (2019). Multidisciplinary approaches for studying rhizobium–legume symbioses. *Canadian Journal of Microbiology, 65*(1), 1–33. https://doi.org/10.1139/cjm-2018-0377

Djamei, A., & Kahmann, R. (2012). Ustilago maydis: Dissecting the molecular interface between pathogen and plant. *PLoS Pathogens, 8*(11), 1–5. https://doi.org/10.1371/journal.ppat.1002955

Dombrecht, B., Gang, P. X., Sprague, S. J., Kirkegaard, J. A., Ross, J. J., Reid, J. B., Fitt, G. P., Sewelam, N., Schenk, P. M., Manners, J. M., & Kazana, K. (2007). MYC2 differentially modulates diverse jasmonate-dependent functions in Arabidopsis. *Plant Cell, 19*(7), 2225–2245. https://doi.org/10.1105/tpc.106.048017

Dong, L., Li, Y., Xu, J., Yang, J., Wei, G., Shen, L., Ding, W., & Chen, S. (2019). Biofertilizers regulate the soil microbial community and enhance *Panax ginseng* yields. *Chinese Medicine (United Kingdom), 14*(1), 1–14. https://doi.org/10.1186/s13020-019-0241-1

Drogue, B., Sanguin, H., Borland, S., Prigent-Combaret, C., & Wisniewski-Dyé, F. (2014). Genome wide profiling of *Azospirillum lipoferum* 4B gene expression during interaction with rice roots. *FEMS Microbiology Ecology, 87*(2), 543–555. https://doi.org/10.1111/1574-6941.12244

Du, G., Zhan, J., Li, J., You, Y., Zhao, Y., & Huang, W. (2012). Effect of fermentation temperature and culture medium on glycerol and ethanol during wine fermentation. *American Journal of Enology and Viticulture, 63*(1), 132–138. https://doi.org/10.5344/ajev.2011.11067

Dutta, S., Yu, S., & Lee, Y. H. (2020). Assessment of the contribution of antagonistic secondary metabolites to the antifungal and biocontrol activities of Pseudomonas fluorescens NBC275. *Plant Pathology Journal, 36*(5), 491–496.

Egamberdieva, D., Wirth, S., Li, L., Abd-Allah, E. F., & Lindström, K. (2017). Microbial cooperation in the rhizosphere improves liquorice growth under salt stress. *Bioengineered, 8*(4), 433–438. https://doi.org/10.1080/21655979.2016.1250983

Ehling-Schulz, M., Koehler, T. M., & Lereclus, D. (2019). The *Bacillus cereus* group: *Bacillus* species with pathogenic potential. *Gram-positive pathogens* (Vol. 7)(3). https://doi.org/10.1128/9781683670131.ch55

Ejigu, G. F., & Jung, J. (2020). Review on the computational genome annotation of sequences obtained by next-generation sequencing. *Biology, 9*(9), 1–27. https://doi.org/10.3390/biology9090295

El-Tarabily, K. A., Soliman, M. H., Nassar, A. H., Al-Hassani, H. A., Sivasithamparam, K., McKenna, F., & Hardy, G. E. S. J. (2000). Biological control of sclerotinia minor using a chitinolytic bacterium and actinomycetes. *Plant Pathology, 49*(5), 573–583. https://doi.org/10.1046/j.1365-3059.2000.00494.x

Elsakhawy, T. A., Fetyan, N. A. H., & Ghazi, A. A. (2019). The potential use of ectoine produced by a moderately halophilic bacteria chromohalobacter salexigens KT989776 for enhancing germination and primary seedling of flax "Linum usitatissimum L." under salinity conditions. *Biotechnology Journal International, 23*(3), 1–12. https://doi.org/10.9734/bji/2019/v23i330078

Escribano-Viana, R., López-Alfaro, I., López, R., Santamaría, P., Gutiérrez, A. R., & González-Arenzana, L. (2018). Impact of chemical and biological fungicides applied to grapevine on grape biofilm, must, and wine microbial diversity. *Frontiers in Microbiology, 9*(59). https://doi.org/10.3389/fmicb.2018.00059

Etalo, D. W., Jeon, J. S., & Raaijmakers, J. M. (2018). Modulation of plant chemistry by beneficial root microbiota. *Natural Product Reports, 35*(5), 398–409. https://doi.org/10.1039/c7np00057j

Etesami, H., & Maheshwari, D. K. (2018). Use of plant growth promoting rhizobacteria (PGPRs) with multiple plant growth promoting traits in stress agriculture: Action mechanisms and future prospects. *Ecotoxicology and Environmental Safety, 156*(January), 225–246. https://doi.org/10.1016/j.ecoenv.2018.03.013

Etesami, H., Alikhani, H. A., & Hosseini, H. M. (2015). Indole-3-acetic acid (IAA) production trait, a useful screening to select endophytic and rhizosphere competent bacteria for rice growth promoting agents. *Methods, 2*, 72–78. https://doi.org/10.1016/j.mex.2015.02.008

Fan, X., Hu, H., Huang, G., Huang, F., Li, Y., & Palta, J. (2015). Soil inoculation with Burkholderia sp. LD-11 has positive effect on water-use efficiency in inbred lines of maize. *Plant and Soil, 390*(1–2), 337–349. https://doi.org/10.1007/s11104-015-2410-z

Fang, L. (2020). *Overview of biofertilizer registration in China. Agrochemical regulatory news and database*. https://agrochemical.chemlinked.com/chempedia/overview-biofertilizer-registration-china.

FAO. (2004). *The State of Food and Agriculture 2003-2004.*

FAO. (2017). Introduction and overview. In En F.a. Organization, & J. Bruinsma (Eds.), *World agriculture: towards 2015/2030: AN FAO perspective* (pp. 1–28). London: Earthscan Publications Ltd.

FAO. (2018). *The future of food and agriculture – alternative pathways to 2050. Summary version.*

Farooq, M. A., Niazi, A. K., Akhtar, J., Saifullah, Farooq, M., Souri, Z., Karimi, N., & Rengel, Z. (2019). Acquiring control: The evolution of ROS-induced oxidative stress and redox signaling pathways in plant stress responses. *Plant Physiology and Biochemistry, 141*, 353–369. https://doi.org/10.1016/j.plaphy.2019.04.039

Fatima, S., & Anjum, T. (2017). Identification of a potential ISR determinant from pseudomonas aeruginosa PM12 against fusarium wilt in tomato. *Frontiers in Plant Science, 8*(May), 1–14. https://doi.org/10.3389/fpls.2017.00848

Federación mundial de Colecciones de Cultivo (FELACC). (2010). *Recomendaciones para el Establecimiento Y funcionamiento de Colecciones de Cultivos de Microorganismos.*

Ferguson, B. J., Indrasumunar, A., Hayashi, S., Lin, M. H., Lin, Y. H., Reid, D. E., & Gresshoff, P. M. (2010). Molecular analysis of legume nodule development and autoregulation. *Journal of Integrative Plant Biology, 52*(1), 61–76. https://doi.org/10.1111/j.1744-7909.2010.00899.x

Fernandes Júnior, P. I., Rohr, T. G., Oliveira, P. J. de, Xavier, G. R., & Rumjanek, N. G. (2009). Polymers as carriers for rhizobial inoculant formulations. *Pesquisa Agropecuária Brasileira, 44*(9), 1184–1190. https://doi.org/10.1590/s0100-204x2009000900017

Fernández, L. A., Zalba, P., Gómez, M. A., & Sagardoy, M. A. (2005). Bacterias solubilizadoras de fosfato inorgánico aisladas de suelos de la región sojera. *Ciencia Del Suelo, 23*(1), 31–37.

Ferraz Helene, L. C., Klepa, M. S., & Hungria, M. (2022). New insights into the taxonomy of bacteria in the genomic era and a case study with rhizobia. *International Journal of Microbiology, 2022*, 4623713.

Ferreira, C. M. H., Soares, H. M. V. M., & Soares, E. V. (2019). Promising bacterial genera for agricultural practices: An insight on plant growth-promoting properties and microbial safety aspects. *Science of the Total Environment, 682*, 779–799. https://doi.org/10.1016/j.scitotenv.2019.04.225

Finn, R. D., Coggill, P., Eberhardt, R. Y., Eddy, S. R., Mistry, J., Mitchell, A. L., Potter, S. C., Punta, M., Qureshi, M., Sangrador-Vegas, A., Salazar, G. A., Tate, J., & Bateman, A. (2016). The Pfam protein families database: Towards a more sustainable future. *Nucleic Acids Research, 44*(D1), D279–D285. https://doi.org/10.1093/nar/gkv1344

Flores, E. R., Perez, F., & De La Torre, M. (1997). Scale-Up of *Bacillus thuringiensis* fermentation based on oxygen transfer. *Journal of Fermentation and Bioengineering, 83*(6), 561–564.

Forouhar, F., Yang, Y., Kumar, D., Chen, Y., Fridman, E., Park, S. W., Chiang, Y., Acton, T. B., Montelione, G. T., Pichersky, E., Klessig, D. F., & Tong, L. (2005). Structural and biochemical studies identify tobacco SABP2 as a methyl salicylate esterase and implicate it in plant innate immunity. *Proceedings of the National Academy of Sciences of the United States of America, 102*(5), 1773–1778. https://doi.org/10.1073/pnas.0409227102

Friesen, T. L., Faris, J. D., Solomon, P. S., & Oliver, R. P. (2008). Host-specific toxins: Effectors of necrotrophic pathogenicity. *Cellular Microbiology, 10*(7), 1421–1428. https://doi.org/10.1111/j.1462-5822.2008.01153.x

Fukase, E., & Martin, W. (2016). Who will feed China in the 21st century? Income growth and food demand and supply in China. *Journal of Agricultural Economics, 67*(1), 3–23. https://doi.org/10.1111/1477-9552.12117

Galaviz, C., Lopez, B. R., De-Bashan, L. E., Hirsch, A. M., Maymon, M., & Bashan, Y. (2018). Root growth improvement of mesquite seedlings and bacterial rhizosphere and soil community changes are induced by inoculation with plant growth-promoting bacteria and promote restoration of eroded desert soil. *Land Degradation and Development Root, 29*(5), 1453–1466. https://doi.org/10.1002/ldr.2904

Galindo, F. S., Teixeira Filho, M. C. M., Buzetti, S., Rodrigues, W. L., Santini, J. M. K., & Alves, C. J. (2019). Nitrogen fertilisation efficiency and wheat grain yield affected by nitrogen doses and sources associated with *Azospirillum brasilense*. *Acta Agriculturae Scandinavica Section B: Soil and Plant Science, 69*(7), 606–617. https://doi.org/10.1080/09064710.2019.1628293

Gilmour, K. A., Davie, C. T., & Gray, N. (2022). Survival and activity of an indigenous iron-reducing microbial community from MX80 bentonite in high temperature /low water environments with relevance to a proposed method of nuclear waste disposal. *Science of the Total Environment, 814*, 152660. https://doi.org/10.1016/j.scitotenv.2021.152660

Glazebrook, J. (2005). Contrasting mechanisms of defense against biotrophic and necrotrophic pathogens. *Annual Review of Phytopathology, 43*, 205–227. https://doi.org/10.1146/annurev.phyto.43.040204.135923

Glick, B. R. (2012). Plant growth-promoting bacteria: Mechanisms and applications. *Scientifica, 2012*(963401).

Glick, B. R. (2020). Beneficial plant-bacterial interactions. In *Beneficial plant-bacterial interactions*. https://doi.org/10.1007/978-3-030-44368-9

Goellner, K., & Conrath, U. (2008). Priming: It's all the world to induced disease resistance. *European Journal of Plant Pathology, 121*(3), 233–242. https://doi.org/10.1007/s10658-007-9251-4

Gold, T. (1992). The deep, hot biosphere. *Proceedings of the National Academy of Sciences of the United States of America, 89*(13), 6045–6049. https://doi.org/10.1073/pnas.89.13.6045

Gómez-Godínez, L. J., Fernandez-Valverde, S. L., Martinez Romero, J. C., & Martínez-Romero, E. (2019). Metatranscriptomics and nitrogen fixation from the rhizoplane of maize plantlets inoculated with a group of PGPRs. *Systematic and Applied Microbiology, 42*(4), 517–525. https://doi.org/10.1016/j.syapm.2019.05.003

Gómez-Godínez, L. J., Martínez-Romero, E., Banuelos, J., & Arteaga-Garibay, R. I. (2021). Tools and challenges to exploit microbial communities in agriculture. *Current Research in Microbial Sciences, 2*. https://doi.org/10.1016/j.crmicr.2021.100062

Gonzalez, E. J., Hernandez, J. P., de-Bashan, L. E., & Bashan, Y. (2018). Dry micro-polymeric inoculant of Azospirillum brasilense is useful for producing mesquite transplants for reforestation of degraded arid zones. *Applied Soil Ecology, 129*(February), 84–93. https://doi.org/10.1016/j.apsoil.2018.04.011

Goris, J., Konstantinidis, K. T., Klappenbach, J. A., Coenye, T., Vandamme, P., & Tiedje, J. M. (2007). DNA-DNA hybridization values and their relationship to whole-genome sequence similarities. *International Journal of Systematic and Evolutionary Microbiology, 57*(1), 81–91. https://doi.org/10.1099/ijs.0.64483-0

Gourgues, M., Brunet-Simon, A., Lebrun, M. H., & Levis, C. (2004). The tetraspanin BcPIs1 is required for appressorium-mediated penetration of *Botrytis cinerea* into host plant leaves. *Molecular Microbiology, 51*(3), 619–629. https://doi.org/10.1046/j.1365-2958.2003.03866.x

Gowen, S., Davies, K. G., & Pembroke, B. (2007). Potential use of pasteuria spp. In *The management of plant parasitic nematodes. Integrated management and biocontrol of vegetable and grain crops nematodes* (pp. 205–219). https://doi.org/10.1007/978-1-4020-6063-2_10

Grageda-Cabrera, O. A., Díaz-Franco, A., Peña-Cabriales, J. J., & Vera-Nuñez, J. A. (2018). Impacto de los biofertilizantes en la agricultura. *Revista Mexicana de Ciencias Agrícolas, 3*(6), 1261–1274. https://doi.org/10.29312/remexca.v3i6.1376

Guijarro, B., Melgarejo, P., Torres, R., Lamarca, N., Usall, J., & De Cal, A. (2007). Effects of different biological formulations of penicillium frequentans on brown rot of peaches. *Biological Control, 42*(1), 86–96. https://doi.org/10.1016/j.biocontrol.2007.03.014

Guimarães, R. A., Pherez-Perrony, P. E., Müller, H., Berg, G., Medeiros, F. H. V., & Cernava, T. (2020). Microbiome-guided evaluation of *Bacillus subtilis* BIOUFLA2 application to reduce mycotoxins in maize kernels. *Biological Control, 150*, 104370. https://doi.org/10.1016/j.biocontrol.2020.104370

Gulati, A., Vyas, P., Rahi, P., & Kasana, R. C. (2009). Plant growth-promoting and rhizosphere-competent acinetobacter rhizosphaerae strain BIHB 723 from the cold deserts of the himalayas. *Current Microbiology, 58*(9), 371–377. https://doi.org/10.1007/s00284-008-9339-x

Gupta, G., Parihar, S. S., Ahirwar, N. K., Snehi, S. K., & Singh, V. (2015). Plant growth promoting rhizobacteria (PGPR): Current and future prospects for development of sustainable agriculture. *Journal of Microbial and Biochemical Technology, 7*(2), 96–102. https://doi.org/10.4172/1948-5948.1000188

Gurevich, A., Saveliev, V., Vyahhi, N., & Tesler, G. (2013). Quast: Quality assessment tool for genome assemblies. *Bioinformatics, 29*(8), 1072–1075. https://doi.org/10.1093/bioinformatics/btt086

Haft, D. H., Selengut, J. D., & White, O. (2003). The TIGRFAMs database of protein families. *Nucleic Acids Research, 31*(1), 371–373. https://doi.org/10.1093/nar/gkg128

Hakim, S., Naqqash, T., Nawaz, M. S., Laraib, I., Siddique, M. J., Zia, R., Mirza, M. S., & Imran, A. (2021). Rhizosphere engineering with plant growth-promoting microorganisms for agriculture and ecological sustainability. *Frontiers in Sustainable Food Systems, 5*(February), 1–23. https://doi.org/10.3389/fsufs.2021.617157

Han, S. W., & Jung, H. W. (2013). Molecular sensors for plant immunity; pattern recognition receptors and race-specific resistance proteins. *Journal of Plant Biology, 56*(6), 357–366. https://doi.org/10.1007/s12374-013-0323-z

Handakumbura, P. P., Rivas Ubach, A., & Battu, A. K. (2021). Visualizing the hidden half: Plant-microbe interactions in the rhizosphere. *MSystems, 6*(5), 1–6. https://doi.org/10.1128/msystems.00765-21

Harbort, C. J., Hashimoto, M., Inoue, H., Niu, Y., Guan, R., Rombolà, A. D., Kopriva, S., Voges, M. J. E. E. E., Sattely, E. S., Garrido-Oter, R., & Schulze-Lefert, P. (2020). Root-Secreted coumarins and the microbiota interact to improve iron nutrition in arabidopsis. *Cell Host and Microbe, 28*(6), 825–837.e6. https://doi.org/10.1016/j.chom.2020.09.006

Hawksworth, D. L., & Kirsop, B. E. (1988). *Living resources for biotechnology: Filamentous fungi.* Cambridge University Press.

Hemetsberger, C., Herrberger, C., Zechmann, B., Hillmer, M., & Doehlemann, G. (2012). The *Ustilago maydis* effector Pep1 suppresses plant immunity by inhibition of host peroxidase activity. *PLoS Pathogens, 8*(5). https://doi.org/10.1371/journal.ppat.1002684

Herridge, D. F. (2008). Inoculation technology for legumes. In M. J. Dilworth, E. K. James, J. I. Sprent, & W. E. Newton (Eds.), *Nitrogen-fixing leguminous symbioses* (pp. 77–115). Springer. https://doi.org/10.1007/978-1-4020-3548-7_4

Herrmann, L., & Lesueur, D. (2013). Challenges of formulation and quality of biofertilizers for successful inoculation. *Applied Microbiology and Biotechnology, 97*(20), 8859–8873. https://doi.org/10.1007/s00253-013-5228-8

Hider, R. C., & Kong, X. (2010). Chemistry and biology of siderophores. *Natural Product Reports, 27*(5), 637–657. https://doi.org/10.1039/b906679a

Hiruma, K., Gerlach, N., Sacristán, S., Nakano, R. T., Hacquard, S., Kracher, B., Neumann, U., Ramírez, D., Bucher, M., O'Connell, R. J., & Schulze-Lefert, P. (2016). Root endophyte colletotrichum tofieldiae confers plant fitness benefits that are phosphate status dependent. *Cell, 165*(2), 464–474. https://doi.org/10.1016/j.cell.2016.02.028

Hodson de Jaramillo, E. (2018). Bioeconomía: El futuro sostenible. *Revista de La Academia Colombiana de Ciencias Exactas, Físicas y Naturales, 42*(164), 188. https://doi.org/10.18257/raccefyn.650

Holt, C., & Yandell, M. (2011). MAKER2: An annotation pipeline and genome-database management tool for second-generation genome projects. *BMC Bioinformatics, 12*(1). https://doi.org/10.1186/1471-2105-12-491

Hong, Y., Zhou, Q., Hao, Y., & Huang, A. C. (2022). Crafting the plant root metabolome for improved microbe-assisted stress resilience. *New Phytologist, 234*(6), 1945–1950. https://doi.org/10.1111/nph.17908

Horvath, P., & Barrangou, R. (2010). CRISPR/Cas, the immune system of Bacteria and Archaea. *Science, 327*(5962), 167–170. https://doi.org/10.1126/science.1179555

Hou, M. P., & Oluranti, B. O. (2013). Evaluation of plant growth promoting potential of four rhizobacterial species for indigenous system. *Journal of Central South University, 20*, 164–171. https://doi.org/10.1007/s11771-013-1472-4

Hou, S., Liu, Z., Shen, H., & Wu, D. (2019). Damage-associated molecular pattern-triggered immunity in plants. *Frontiers in Plant Science, 10*(May). https://doi.org/10.3389/fpls.2019.00646

Houterman, P. M., Cornelissen, B. J. C., & Rep, M. (2008). Suppression of plant resistance gene-based immunity by a fungal effector. *PLoS Pathogens, 4*(5), 1–6. https://doi.org/10.1371/journal.ppat.1000061

Hu, J., Wei, Z., Friman, V.-P. P., Gu, S. H., Wang, X. F., Eisenhauer, N., Yang, T. J., Ma, J., Shen, Q. R., Xu, Y. C., Jousseta, A., & Jousset, A. (2016). Probiotic diversity enhances rhizosphere microbiome function and plant disease suppression. . *Mbio, 7*(6), 1–8. https://doi.org/10.1128/mBio.01790-16

Hu, S., Li, K., Zhang, Y., Wang, Y., Fu, L., Xiao, Y., Tang, X., & Gao, J. (2022). New insights into the threshold values of multi-locus sequence analysis, average nucleotide identity and digital DNA–DNA hybridization in delineating streptomyces species. *Frontiers in Microbiology, 13*(May), 1–10. https://doi.org/10.3389/fmicb.2022.910277

Hudzik, C., Hou, Y., Ma, W., & Axtell, M. J. (2020). Exchange of small regulatory rnas between plants and their pests. *Plant Physiology, 182*(1), 51–62. https://doi.org/10.1104/pp.19.00931

Hung, R., & Lee Rutgers, S. (2016). Applications of *Aspergillus* in plant growth promotion. In V. K. Gupta (Ed.), *New and future developments in microbial biotechnology and bioengineering* (pp. 223–227). Elsevier B.V. https://doi.org/10.1016/B978-0-444-63505-1.00018-X

Hussain, T., Akthar, N., Aminedi, R., Danish, M., Nishat, Y., & Patel, S. (2020). Role of the potent microbial based bioagents and their emerging strategies for the ecofriendly management of agricultural phytopathogens. In Joginder Singh, & A. N. Yadav (Eds.), *Natural bioactive products in sustainable agriculture* (pp. 45–66). Springer. https://doi.org/10.1007/978-981-15-3024-1_4

Iavicoli, A., Boutet, E., Buchala, A., & Métraux, J. P. (2003). Induced systemic resistance in *Arabidopsis thaliana* in response to root inoculation with pseudomonas fluorescens CHA0. *Molecular Plant-Microbe Interactions, 16*(10), 851–858. https://doi.org/10.1094/MPMI.2003.16.10.851

Ibarra-Villarreal, A. L., Gándara-Ledezma, A., Godoy-Flores, A. D., Herrera-Sepúlveda, A., Díaz-Rodríguez, A. M., Parra-Cota, F. I., & de los Santos-Villalobos, S. (2021). Salt-tolerant *Bacillus* species as a promising strategy to mitigate the salinity stress in wheat (*Triticum turgidum* subsp. durum). *Journal of Arid Environments, 186*. https://doi.org/10.1016/j.jaridenv.2020.104399

Ibrahim, M. E. (2017). Trials on the application of fertilization combined with plant hormone spraying for improving the production of carnation absolute oil. *Journal of Materials and Environmental Science, 8*(4), 1284–1290.

Idnurm, A., & Howlett, B. J. (2001). Pathogenicity genes of phytopathogenic fungi. *Molecular Plant Pathology, 2*(4), 241–255. https://doi.org/10.1046/j.1464-6722.2001.00070.x

Idris, E. S. E., Iglesias, D. J., Talon, M., & Borriss, R. (2007). Tryptophan-dependent production of Indole-3-Acetic Acid (IAA) affects level of plant growth promotion by *Bacillus amyloliquefaciens* FZB42. *Molecular Plant-Microbe Interactions, 20*(6), 619–626. https://doi.org/10.1094/MPMI-20-6-0619

Ijaq, J., Chandrasekharan, M., Poddar, R., Bethi, N., & Sundararajan, V. S. (2015). Annotation and curation of uncharacterized proteins- challenges. *Frontiers in Genetics, 6*(March), 1–7. https://doi.org/10.3389/fgene.2015.00119

Imperiali, N., Dennert, F., Schneider, J., Laessle, T., Velatta, C., Fesselet, M., Wyler, M., Mascher, F., Mavrodi, O., Mavrodi, D., Maurhofer, M., & Keel, C. (2017). Relationships between root pathogen resistance, abundance and expression of pseudomonas antimicrobial genes, and soil properties in representative swiss agricultural soils. *Frontiers in Plant Science, 8*(March), 1–22. https://doi.org/10.3389/fpls.2017.00427

Ininbergs, K., Bay, G., Rasmussen, U., Wardle, D. A., & Nilsson, M. C. (2011). Composition and diversity of nifH genes of nitrogen-fixing cyanobacteria associated with boreal forest feather mosses. *New Phytologist, 192*(2), 507–517. https://doi.org/10.1111/j.1469-8137.2011.03809.x

Iriti, M., Scarafoni, A., Pierce, S., Castorina, G., & Vitalini, S. (2019). Soil application of effective microorganisms (EM) maintains leaf photosynthetic efficiency, increases seed yield and quality traits of bean (*Phaseolus vulgaris* L.) plants grown on different substrates. *International Journal of Molecular Sciences, 20*(9). https://doi.org/10.3390/ijms20092327

Islam, W., Noman, A., Naveed, H., Huang, Z., & Chen, H. Y. H. (2020). Role of environmental factors in shaping the soil microbiome. *Environmental Science and Pollution Research, 27*(33), 41225–41247. https://doi.org/10.1007/s11356-020-10471-2

Jaemsaeng, R., Jantasuriyarat, C., & Thamchaipenet, A. (2018). Molecular interaction of 1-aminocyclopropane-1-carboxylate deaminase (ACCD)-producing endophytic Streptomyces sp. GMKU 336 towards salt-stress resistance of *Oryza sativa* L. cv. KDML105. *Scientific Reports, 8*(1), 1–15. https://doi.org/10.1038/s41598-018-19799-9

Jankele, R., & Svoboda, P. (2014). TAL effectors: Tools for DNA targeting. *Briefings in Functional Genomics, 13*(5), 409–419. https://doi.org/10.1093/bfgp/elu013

Jha, C. K., & Saraf, M. (2012). Evaluation of multispecies plant-growth-promoting consortia for the growth promotion of *Jatropha curcas* L. *Journal of Plant Growth Regulation, 31*(4), 588–598. https://doi.org/10.1007/s00344-012-9269-5

Jiang, C., Fan, Z., Li, Z., Niu, D., Li, Y., Zheng, M., Wang, Q., Jin, H., & Guo, J. (2020). *Bacillus cereus* AR156 triggers induced systemic resistance against *Pseudomonas syringae* pv. tomato DC3000 by suppressing miR472 and activating CNLs-mediated basal immunity in Arabidopsis. *Molecular Plant Pathology, 21*(6), 854–870. https://doi.org/10.1111/mpp.12935

Johnson, J. L., & Whitman, W. B. (2007). Similarity analysis of DNA. In C. A. Reddy (Ed.), *Methods for general and molecular microbiology* (pp. 624–652).

Johnson, R. D., Johnson, L., Itoh, Y., Kodama, M., Otani, H., & Kohmoto, K. (2000). Cloning and characterization of a cyclic peptide synthetase gene from *Alternaria alternata* apple pathotype whose product is involved in AM-toxin synthesis and pathogenicity. *Molecular Plant-Microbe Interactions, 13*(7), 742–753. https://doi.org/10.1094/MPMI.2000.13.7.742

Jupe, J., Stam, R., Howden, A. J. M., Morris, J. A., Zhang, R., Hedley, P. E., & Huitema, E. (2013). Phytophthora capsici-tomato interaction features dramatic shifts in gene expression associated with a hemi-biotrophic lifestyle. *Genome Biology, 14*(6). https://doi.org/10.1186/gb-2013-14-6-r63

van Kan, J. A. L. (2006). Licensed to kill: The lifestyle of a necrotrophic plant pathogen. *Trends in Plant Science, 11*(5), 247–253. https://doi.org/10.1016/j.tplants.2006.03.005

Kamilova, F., Okon, Y., & Katja Hora, Sandra deWeert (2015). Commercialization of microbes: Manufacturing, inoculation, best practice for objective field testing, and registration. In B. Lugtenberg (Ed.), *Principles of plant-microbe interactions: Microbes for sustainable agriculture* (pp. 1–448). https://doi.org/10.1007/978-3-319-08575-3

Kang, W., Zhu, X., Wang, Y., Chen, L., & Duan, Y. (2018). Transcriptomic and metabolomic analyses reveal that bacteria promote plant defense during infection of soybean cyst nematode in soybean. *BMC Plant Biology, 18*(1), 1–14. https://doi.org/10.1186/s12870-018-1302-9

Kang, A., Zhang, N., Xun, W., Dong, X., Xiao, M., Liu, Z., Xu, Z., Feng, H., Zou, J., Shen, Q., & Zhang, R. (2022). Nitrogen fertilization modulates beneficial rhizosphere interactions through signaling effect of nitric oxide. *Plant Physiology, 188*(2), 1129–1140. https://doi.org/10.1093/plphys/kiab555

Kapulnik, Y., & Koltai, H. (2016). Fine-tuning by strigolactones of root response to low phosphate. *Journal of Integrative Plant Biology, 58*(3), 203–212. https://doi.org/10.1111/jipb.12454

Kars, I., Krooshof, G. H., Wagemakers, L., Joosten, R., Benen, J. A. E., & Van Kan, J. A. L. (2005). Necrotizing activity of five *Botrytis cinerea* endopolygalacturonases produced in Pichia pastoris. *Plant Journal, 43*(2), 213–225. https://doi.org/10.1111/j.1365-313X.2005.02436.x

Kaur, S., Egidi, E., Qiu, Z., Macdonald, C. A., Verma, J. P., Trivedi, P., Wang, J., Liu, H., & Singh, B. K. (2022). Synthetic community improves crop performance and alters rhizosphere microbial communities. *Journal of Sustainable Agriculture and Environment, 1*(2), 118–131. https://doi.org/10.1002/sae2.12017

Kaushik, S., & Djiwanti, S. R. (2019). Nanofertilizers: Smart delivery of plant nutrients. In D. P, & Y. J (Eds.), *Nanotechnology for agriculture: Crop production and protection* (pp. 59–72). Springer. https://doi.org/10.1007/978-981-32-9374-8_3

Keen, N. T. (1990). *Plant -pathogen interactions.*

Kennedy, I. R., Choudhury, A. T. M. A., Kecskés, M. L., & Rose, M. T. (2008). Efficient nutrient use in rice production in Vietnam achieved using inoculant biofertilisers. In I. R. Kennedy, A. T. M. A. Choudhury, M. L. Kecskés, & M. T. Rose (Eds.), *Proceedings of a project (SMCN/2002/073) workshop held in Hanoi, vietnam, 12–13 October 2007* (AciAr Proc).

Kent, W. J. (2002). Blat —the BLAST -like alignment tool. *Genome Research, 12*(4), 656–664. https://doi.org/10.1101/gr.229202

Khan, N., Bano, A., & Curá, J. A. (2020). Role of beneficial microorganisms and salicylic acid in improving rainfed agriculture and future food safety. *Microorganisms, 8*(7), 1–22. https://doi.org/10.3390/microorganisms8071018

Khare, E., & Arora, N. K. (2015). Effects of soil environment on field efficacy of microbial inoculants. In *Plant microbes symbiosis: Applied facets* (pp. 353–381). https://doi.org/10.1007/978-81-322-2068-8

Khare, E., Kumar, S., & Kim, K. (2018). Role of peptaibols and lytic enzymes of *Trichoderma cerinum* Gur1 in biocontrol of fusraium oxysporum and chickpea wilt. *Environmental Sustainability, 1*(1), 39–47. https://doi.org/10.1007/s42398-018-0001-7

Khatabi, B., Gharechahi, J., Ghaffari, M. R., Liu, D., Haynes, P. A., McKay, M. J., Mirzaei, M., & Salekdeh, G. H. (2019). Plant–microbe symbiosis: What has proteomics taught us? *Proteomics, 19*(16), 1–36. https://doi.org/10.1002/pmic.201800105

Khurana, A., & Kumar, V. (2022). *State of biofertilizers and organic fertilizers in India.* Centre for Science and Environment.

Khurana, S. M. P., Pandey, S. K., Sarkar, D., & Chanemougasoundharam, A. (2005). Apoptosis in plant disease response: A close encounter of the pathogen kind. *Current Science, 88*(5), 740–752.

Kierul, K., Voigt, B., Albrecht, D., Chen, X. H., Carvalhais, L. C., & Borriss, R. (2015). Influence of root exudates on the extracellular proteome of the plant growth-promoting bacterium bacillus amyloliquefaciens FZB42. *Microbiology (United Kingdom), 161*(1), 131–147. https://doi.org/10.1099/mic.0.083576-0

Kim, M., & Chun, J. (2014). 16S rRNA gene-based identification of bacteria and archaea using the EzTaxon server. In *Methods in microbiology* (1st ed., Vol. 41). Elsevier Ltd. https://doi.org/10.1016/bs.mim.2014.08.001

Kim, M., Oh, H. S., Park, S. C., & Chun, J. (2014). Towards a taxonomic coherence between average nucleotide identity and 16S rRNA gene sequence similarity for species demarcation of prokaryotes. *International Journal of Systematic and Evolutionary Microbiology, 64*(PART 2), 346–351. https://doi.org/10.1099/ijs.0.059774-0

Kirsop, B. E., & DaSilva, E. J. (1988). Organization of resource centers. In B. E. Kirsop, & C. P. Kurtzman (Eds.), *Living resources for biotechnology: Yeasts* (pp. 173–187). Cambridge University Press.

Kirsop, B. E., & Kurtzman, C. P. (1988). *Living resources for biotechnology: Yeasts.* Cambridge University Press.

Kirsop, B. E. (1988). Resources centres. In B. E. Kirsop, & C. P. Kurtzman (Eds.), *Living resources for biotechnology: Yeasts* (pp. 1–35). Cambridge University Press.

Klonowska, A., Melkonian, R., Miché, L., Tisseyre, P., & Moulin, L. (2018). Transcriptomic profiling of *Burkholderia phymatum* STM815, *Cupriavidus taiwanensis* LMG19424 and *Rhizobium mesoamericanum* STM3625 in response to *Mimosa pudica* root exudates illuminates the molecular basis of their nodulation competitiveness and symbiotic ev. *BMC Genomics, 19*(1), 1–22. https://doi.org/10.1186/s12864-018-4487-2

Knoche, M., Khanal, B. P., & Stopar, M. (2011). Russeting and microcracking of "Golden Delicious" apple fruit concomitantly decline due to gibberellin a4+7 application. *Journal of the American Society for Horticultural Science, 136*(3), 159−164. https://doi.org/10.21273/jashs.136.3.159

Kobayashi, D. Y., & Crouch, J. A. (2009). *Bacterial /fungal interactions: From pathogens to mutualistic endosymbionts*. https://doi.org/10.1146/annurev-phyto-080508-081729

Köhl, J., Kolnaar, R., & Ravensberg, W. J. (2019). *Mode of action of microbial biological control agents against plant diseases: Relevance beyond efficacy ivorlanov*. https://doi.org/10.3389/fpls.2019.00845

Köberl, M., Wagner, P., Müller, H., Matzer, R., Unterfrauner, H., Cernava, T., & Berg, G. (2020). Unraveling the complexity of soil microbiomes in a large-scale study subjected to different agricultural management in styria. *Frontiers in Microbiology, 11*(May), 1−11. https://doi.org/10.3389/fmicb.2020.01052

Köhl, J., Postma, J., Nicot, P., Ruocco, M., & Blum, B. (2011). Stepwise screening of microorganisms for commercial use in biological control of plant-pathogenic fungi and bacteria. *Biological Control, 57*(1), 1−12. https://doi.org/10.1016/j.biocontrol.2010.12.004

Kong, Z., & Liu, H. (2022). Modification of rhizosphere microbial communities: A possible mechanism of plant growth promoting rhizobacteria enhancing plant growth and fitness. *Frontiers in Plant Science, 13*. https://doi.org/10.3389/fpls.2022.920813

Kong, J., Huh, S., Won, J. I., Yoon, J., Kim, B., & Kim, K. (2019). Gaap: A genome assembly + annotation pipeline. *BioMed Research International*. . https://doi.org/10.1155/2019/4767354, 2019.

Koo, Y. M., Heo, A. Y., & Choi, H. W. (2020). Salicylic acid as a safe plant protector and growth regulator. *Plant Pathology Journal, 36*(1), 1−10. https://doi.org/10.5423/PPJ.RW.12.2019.0295

Koprivova, A., Schuck, S., Jacoby, R. P., Klinkhammer, I., Welter, B., Leson, L., Martyn, A., Nauen, J., Grabenhorst, N., Mandelkow, J. F., Zuccaro, A., Zeier, J., & Kopriva, S. (2019). Root-specific camalexin biosynthesis controls the plant growth-promoting effects of multiple bacterial strains. *Proceedings of the National Academy of Sciences of the United States of America, 116*(31), 15735−15744. https://doi.org/10.1073/pnas.1818604116

Koren, S., Walenz, B. P., Berlin, K., Miller, J. R., Bergman, N. H., & Phillippy, A. M. (2017). Canu: Scalable and accurate long-read assembly via adaptive κ-mer weighting and repeat separation. *Genome Research, 27*(5), 722−736. https://doi.org/10.1101/gr.215087.116

Koskey, G., Mburu, S. W., Awino, R., Njeru, E. M., & Maingi, J. M. (2021). Potential use of beneficial microorganisms for soil amelioration, phytopathogen biocontrol, and sustainable crop production in smallholder agroecosystems. *Frontiers in Sustainable Food Systems, 5*(April), 1−20. https://doi.org/10.3389/fsufs.2021.606308

Kovtun, Y., Chiu, W. L., Tena, G., & Sheen, J. (2000). Functional analysis of oxidative stress-activated mitogen-activated protein kinase cascade in plants. *Proceedings of the National Academy of Sciences of the United States of America, 97*(6), 2940−2945. https://doi.org/10.1073/pnas.97.6.2940

Krishnen, G., Kecskés, M. L., Rose, M. T., Geelan-Small, P., Amprayn, K. O., Pereg, L., & Kennedy, I. R. (2011). Field monitoring of plant-growth-promoting rhizobacteria by colony immunoblotting. *Canadian Journal of Microbiology, 57*(11), 914−922. https://doi.org/10.1139/w11-059

Kudjordjie, E. N., Sapkota, R., Steffensen, S. K., Fomsgaard, I. S., & Nicolaisen, M. (2019). Maize synthesized benzoxazinoids affect the host associated microbiome. *Microbiome, 7*(1), 1−17. https://doi.org/10.1186/s40168-019-0677-7

Kügler, S., Cooper, R. E., Boessneck, J., Küsel, K., & Wichard, T. (2020). Rhizobactin B is the preferred siderophore by a novel pseudomonas isolate to obtain iron from dissolved organic matter in peatlands. *BioMetals, 33*(6), 415–433. https://doi.org/10.1007/s10534-020-00258-w

Kumar, B., Trivedi, P., & Pandey, A. (2007). Pseudomonas corrugata: A suitable bacterial inoculant for maize grown under rainfed conditions of himalayan region. *Soil Biology and Biochemistry, 39*, 3093–3100. https://doi.org/10.1016/j.soilbio.2007.07.003

Kumar, M., Singh, D. P., Prabha, R., Rai, A. K., & L.S.. (2016). Role of microbial inoculants in nutrient use efficiency. In D. P. Singh, H. B. Singh, & R. Prabha (Eds.), *Microbial inoculants in sustainable agricultural productivity: Vol. 2: Functional applications* (pp. 1–308). Springer. https://doi.org/10.1007/978-81-322-2644-4

Kurtz, S., Phillippy, A., Delcher, A. L., Smoot, M., Shumway, M., Antonescu, C., & Salzberg, S. L. (2004). Versatile and open software for comparing large genomes. *Genome Biology, 5*(2). https://doi.org/10.1186/gb-2004-5-2-r12

Kusstatscher, P., Wicaksono, W. A., Thenappan, D. P., Adam, E., Müller, H., & Berg, G. (2020). Microbiome management by biological and chemical treatments in maize is linked to plant health. *Microorganisms, 8*(10), 1–14. https://doi.org/10.3390/microorganisms8101506

Kviat, H., Awafo, V., Bender, J., Carter, J., Conway, R., Egli, S., Feeser, T., Jornitz, M., Kearns, M., Levy, R., Madsen, R., Martin, J., McBurnie, L., Meissner, L., Meltzer, T., Pawar, V., Phelan, M., Stinavage, P., Sweeney, N., et al. (2008). Technical report No. 26 revised 2008 sterilizing filtration of liquids PDA. *Journal of Pharmaceutical Science and Technology, 62*(S-5).

L'Hoir, M., & Duponnois, R. (2021). Combining the seed endophytic bacteria and the back to the future approaches for plant holonbiont breeding. *Frontiers in Agronomy, 3*(September), 1–17. https://doi.org/10.3389/fagro.2021.724450

Lakshmanan, V., Castaneda, R., Rudrappa, T., & Bais, H. P. (2013). Root transcriptome analysis of *Arabidopsis thaliana* exposed to beneficial Bacillus subtilis FB17 rhizobacteria revealed genes for bacterial recruitment and plant defense independent of malate efflux. *Planta, 238*(4), 657–668. https://doi.org/10.1007/s00425-013-1920-2

Lamb, C., & Dixon, R. A. (1997). The oxidative burst in plant disease resistance. *Annual Review of Plant Biology, 48*, 251–275. https://doi.org/10.1146/annurev.arplant.48.1.251

Lamour, K. H., Stam, R., Jupe, J., & Huitema, E. (2012). The oomycete broad-host-range pathogen *Phytophthora capsici. Molecular Plant Pathology, 13*(4), 329–337. https://doi.org/10.1111/j.1364-3703.2011.00754.x

Land, M., Hauser, L., Jun, S. R., Nookaew, I., Leuze, M. R., Ahn, T. H., Karpinets, T., Lund, O., Kora, G., Wassenaar, T., Poudel, S., & Ussery, D. W. (2015). Insights from 20 years of bacterial genome sequencing. *Functional and Integrative Genomics, 15*(2), 141–161. https://doi.org/10.1007/s10142-015-0433-4

Lares-Orozco, M. F., Robles-Morúa, A., Yepez, E. A., & Handler, R. M. (2016). Global warming potential of intensive wheat production in the Yaqui valley, Mexico: A resource for the design of localized mitigation strategies. *Journal of Cleaner Production, 127*, 522–532. https://doi.org/10.1016/j.jclepro.2016.03.128

Laveilhé, A., Fochesato, S., Lalaouna, D., Heulin, T., & Achouak, W. (2022). Phytobeneficial traits of rhizobacteria under the control of multiple molecular dialogues. *Microbial Biotechnology, 15*(7), 2083–2096. https://doi.org/10.1111/1751-7915.14023

Lee, I., Kim, Y. O., Park, S. C., & Chun, J. (2016). OrthoANI: An improved algorithm and software for calculating average nucleotide identity. *International Journal of Systematic and Evolutionary Microbiology, 66*(2), 1100–1103. https://doi.org/10.1099/ijsem.0.000760

Lee, D. H., Lal, N. K., Lin, Z. J. D., Ma, S., Liu, J., Castro, B., Toruño, T., Dinesh-Kumar, S. P., & Coaker, G. (2020). Regulation of reactive oxygen species during plant immunity through phosphorylation and ubiquitination of RBOHD. *Nature Communications, 11*(1). https://doi.org/10.1038/s41467-020-15601-5

Leon-Reyes, A., Du, Y., Koornneef, A., Proietti, S., Körbes, A. P., Memelink, J., Pieterse, C. M. J., & Ritsema, T. (2010). Ethylene signaling renders the jasmonate response of arabidopsis insensitive to future suppression by salicylic acid. *Molecular Plant-Microbe Interactions, 23*(2), 187–197. https://doi.org/10.1094/MPMI-23-2-0187

Li, Y., & Chen, S. (2019). Fusaricidin produced by *Paenibacillus polymyxa* WLY78 induces systemic resistance against fusarium wilt of cucumber. *International Journal of Molecular Sciences, 20*(20). https://doi.org/10.3390/ijms20205240

Li, R., Zhu, H., Ruan, J., Qian, W., Fang, X., Shi, Z., Li, Y., Li, S., Shan, G., Kristiansen, K., Li, S., Yang, H., Wang, J., & Wang, J. (2010). De novo assembly of human genomes with massively parallel short read sequencing. *Genome Research, 20*(2), 265–272. https://doi.org/10.1101/gr.097261.109

Li, S., Jochum, C. C., Yu, F., Du, L., Harris, S. D., & Yuen, G. Y. (2008). An antibiotic complex from lysobacter enzymogenes strain C3: Antimicrobial activity and role in plant disease control. *Phytopathology, 98*(6), 695–701.

Li, T., Liu, B., Spalding, M. H., Weeks, D. P., & Yang, B. (2012). High-efficiency TALEN-based gene editing produces disease-resistant rice. *Nature Biotechnology, 30*(5), 390–392. https://doi.org/10.1038/nbt.2199

Li, Y., Li, Q., Guan, G., & Chen, S. (2020). Phosphate solubilizing bacteria stimulate wheat rhizosphere and endosphere biological nitrogen fixation by improving phosphorus content. *PeerJ, 2020*(3). https://doi.org/10.7717/peerj.9062

Li, S. M., Zheng, H. X., Zhang, X. S., & Sui, N. (2021). Cytokinins as central regulators during plant growth and stress response. *Plant Cell Reports, 40*(2), 271–282. https://doi.org/10.1007/s00299-020-02612-1

Lim, H. S., Kim, Y. S., & Kim, S. D. (1991). Pseudomonas stutzeri YPL-1 genetic transformation and antifungal mechanism against *Fusarium solani*, an agent of plant root rot. *Applied and Environmental Microbiology, 57*(2), 510–516. https://doi.org/10.1128/aem.57.2.510-516.1991

Liu, H., & Brettell, L. E. (2019). Plant defense by VOC-induced microbial priming. *Trends in Plant Science, 24*(3), 187–189. https://doi.org/10.1016/j.tplants.2019.01.008

Liu, F., Xing, S., Ma, H., Du, Z., & Ma, B. (2013). Cytokinin-producing, plant growth-promoting rhizobacteria that confer resistance to drought stress in *Platycladus orientalis* container seedlings. *Applied Microbiology and Biotechnology, 97*(20), 9155–9164. https://doi.org/10.1007/s00253-013-5193-2

Liu, Y., Lai, Q., Göker, M., Meier-Kolthoff, J. P., Wang, M., Sun, Y., Wang, L., & Shao, Z. (2015). Genomic insights into the taxonomic status of the *Bacillus cereus* group. *Scientific Reports, 5*, 1–11. https://doi.org/10.1038/srep14082

Liu, Y., Chen, L., Zhang, N., Li, Z., Zhang, G., Xu, Y., Shen, Q., & Zhang, R. (2016). Plant-microbe communication enhances auxin biosynthesis by a root-associated bacterium, *Bacillus amyloliquefaciens* SQR9. *Molecular Plant-Microbe Interactions, 29*(4), 324–330. https://doi.org/10.1094/MPMI-10-15-0239-R

Liu, S., Hao, H., Lu, X., Zhao, X., Wang, Y., Zhang, Y., Xie, Z., & Wang, R. (2017). Transcriptome profiling of genes involved in induced systemic salt tolerance conferred by *Bacillus amyloliquefaciens* FZB42 in *Arabidopsis thaliana*. *Scientific Reports, 7*(1), 1–13. https://doi.org/10.1038/s41598-017-11308-8

Liu, Y., Du, J., Lai, Q., Zeng, R., Ye, D., Xu, J., & Shao, Z. (2017). Proposal of nine novel species of the *Bacillus cereus* group. *International Journal of Systematic and Evolutionary Microbiology, 67*(8), 2499–2508. https://doi.org/10.1099/ijsem.0.001821

Liu, H., Brettell, L. E., & Singh, B. (2020). Linking the phyllosphere microbiome to plant health. *Trends in Plant Science, 25*(9), 841–844. https://doi.org/10.1016/j.tplants.2020.06.003

Liu, H., Li, Y., Ge, K., Du, B., Liu, K., Wang, C., & Ding, Y. (2021). Interactional mechanisms of *Paenibacillus polymyxa* SC2 and pepper (*Capsicum annuum* L.) suggested by transcriptomics. *BMC Microbiology, 21*(1), 1–16. https://doi.org/10.1186/s12866-021-02132-2

Liu, S., Li, H., Daigger, G. T., Huang, J., & Song, G. (2022). Material biosynthesis, mechanism regulation and resource recycling of biomass and high-value substances from wastewater treatment by photosynthetic bacteria: A review. *Science of the Total Environment, 820*, 153200. https://doi.org/10.1016/j.scitotenv.2022.153200

Lobb, B., Tremblay, B. J. M., Moreno-Hagelsieb, G., & Doxey, A. C. (2020). An assessment of genome annotation coverage across the bacterial tree of life. *Microbial Genomics, 6*(3). https://doi.org/10.1099/mgen.0.000341

Lobo, C. B., Juárez Tomás, M. S., Viruel, E., Ferrero, M. A., & Lucca, M. E. (2019). Development of low-cost formulations of plant growth-promoting bacteria to be used as inoculants in beneficial agricultural technologies. *Microbiological Research, 219*(July), 12–25. https://doi.org/10.1016/j.micres.2018.10.012

Lorenzo, O., Piqueras, R., Sánchez-Serrano, J. J., & Solano, R. (2003). Ethylene response FACTOR1 integrates signals from ethylene and jasmonate pathways in plant defense. *Plant Cell, 15*(1), 165–178. https://doi.org/10.1105/tpc.007468

Lu, H., Wu, Z., Wang, W., Xu, X., & Liu, X. (2020). Rs-198 liquid biofertilizers affect microbial community diversity and enzyme activities and promote *Vitis vinifera* L. Growth. *BioMed Research International, 2020*(8321462). https://doi.org/10.1155/2020/8321462

Lugtenberg, B., & Kamilova, F. (2009). Plant-growth-promoting rhizobacteria. *Annual Review of Microbiology, 63*, 541–556. https://doi.org/10.1146/annurev.micro.62.081307.162918

Ma, W., Smigel, A., Tsai, Y. C., Braam, J., & Berkowitz, G. A. (2008). Innate immunity signaling: Cytosolic Ca^{2+} elevation is linked to downstream nitric oxide generation through the action of calmodulin or a calmodulin-like protein. *Plant Physiology, 148*(2), 818–828. https://doi.org/10.1104/pp.108.125104

Madhaiyan, M., Peng, N., Te, N. S., Hsin I, C., Lin, C., Lin, F., Reddy, C., Yan, H., & Ji, L. (2013). Improvement of plant growth and seed yield in *Jatropha curcas* by a novel nitrogen-fixing root associated *Enterobacter* species. *Biotechnology for Biofuels, 6*(1), 1–13. https://doi.org/10.1186/1754-6834-6-140

Magotra, S., Trakroo, D., Ganjoo, S., & Vakhlu, J. (2016). Bacillus -mediated-induced systemic resistance (ISR) against fusarium corm rot. In D. K. Choudhary, & A. Varma (Eds.), *Microbial-mediated induced systemic resistance in plants* (pp. 15–22). Springer.

Mahamdallie, S., Ruark, E., Yost, S., Münz, M., Renwick, A., Poyastro-Pearson, E., Strydom, A., Seal, S., & Rahman, N. (2018). The quality sequencing minimum (QSM): Providing comprehensive, consistent, transparent next generation sequencing data quality assurance [version 1; referees: 2 approved, 1 approved with reservations]. *Wellcome Open Research, 3*(0), 1–10. https://doi.org/10.12688/wellcomeopenres.14307.1

Mahanty, T., Bhattacharjee, S., Das, B., Goswami, M., Bhattacharyya, P., Tribedi, P., & Ghosh, A. (2017). Biofertilizers: A potential approach for sustainable agriculture development. *Environmental Science and Pollution Research, 24*(4). https://doi.org/10.1007/s11356-016-8104-0

Maksimov, I. V., Abizgil'dina, R. R., & Pusenkova, L. I. (2011). Plant growth promoting rhizobacteria as alternative to chemical crop protectors from pathogens (review). *Applied Biochemistry and Microbiology, 47*(4), 333–345. https://doi.org/10.1134/S0003683811040090

Malik, N. A. A., Kumar, I. S., & Nadarajah, K. (2020). Elicitor and receptor molecules: Orchestrators of plant defense and immunity. *International Journal of Molecular Sciences, 21*(3). https://doi.org/10.3390/ijms21030963

Mall, R. K., Gupta, A., & Sonkar, G. (2017). Effect of climate change on agricultural crops. In S. K. Dubey, A. Pandey, & R. S. Sangwan (Eds.), *Current developments in biotechnology and bioengineering: Crop modification, nutrition, and food production* (pp. 23–46). Elsevier B.V. https://doi.org/10.1016/B978-0-444-63661-4.00002-5

Malusá, E., & Vassilev, N. (2014). A contribution to set a legal framework for biofertilisers. *Applied Microbiology and Biotechnology, 98*(15), 6599–6607. https://doi.org/10.1007/s00253-014-5828-y

Malusá, E., Sas-Paszt, L., & Ciesielska, J. (2012). Technologies for beneficial microorganisms inocula used as biofertilizers. *The Scientific World Journal*. . https://doi.org/10.1100/2012/491206, 2012.

Malusá, E., Pinzari, F., & Canfora, L. (2016). Efficacy of biofertilizers: Challenges to improve crop production. In D. P. Singh, H. B. Singh, & R. Prabha (Eds.), *Microbial inoculants in sustainable agricultural productivity: Vol. 2: Functional applications* (pp. 17–40). https://doi.org/10.1007/978-81-322-2644-4

Manfredini, A., Malusà, E., Costa, C., Pallottino, F., Mocali, S., Pinzari, F., & Canfora, L. (2021). Current methods, common practices, and perspectives in tracking and monitoring bioinoculants in soil. *Frontiers in Microbiology, 12*(August), 1–22. https://doi.org/10.3389/fmicb.2021.698491

Manikandan, R., Saravanakumar, D., Rajendran, L., Raguchander, T., & Samiyappan, R. (2010). Standardization of liquid formulation of pseudomonas fluorescens Pf1 for its efficacy against fusarium wilt of tomato. *Biological Control, 54*(2), 83–89. https://doi.org/10.1016/j.biocontrol.2010.04.004

Manzotti, A., Bergna, A., Burow, M., Jørgensen, H. J. L., Cernava, T., Berg, G., Collinge, D. B., & Jensen, B. (2020). Insights into the community structure and lifestyle of the fungal root endophytes of tomato by combining amplicon sequencing and isolation approaches with phytohormone profiling. *FEMS Microbiology Ecology, 96*, 103–111.

Marchler-Bauer, A., Lu, S., Anderson, J. B., Chitsaz, F., Derbyshire, M. K., DeWeese-Scott, C., Fong, J. H., Geer, L. Y., Geer, R. C., Gonzales, N. R., Gwadz, M., Hurwitz, D. I., Jackson, J. D., Ke, Z., Lanczycki, C. J., Lu, F., Marchler, G. H., Mullokandov, M., Omelchenko, M. V., et al. (2011). Cdd: A conserved domain database for the functional annotation of proteins. *Nucleic Acids Research, 39*(1), 225–229. https://doi.org/10.1093/nar/gkq1189

Margulies, M., Egholm, M., Altman, W. E., Attiya, S., Bader, J. S., Bemben, L. A., Berka, J., Braverman, M. S., Chen, Y. J., Chen, Z., Dewell, S. B., Du, L., Fierro, J. M., Gomes, X. V., Godwin, B. C., He, W., Helgesen, S., Ho, C. H., Irzyk, G. P., et al. (2005). Genome sequencing in microfabricated high-density picolitre reactors. *Nature, 437*(7057), 376–380. https://doi.org/10.1038/nature03959

Martínez-Granero, F., Rivilla, R., & Martín, M. (2006). Rhizosphere selection of highly motile phenotypic variants of pseudomonas fluorescens with enhanced competitive colonization ability. *Applied and Environmental Microbiology, 72*(5), 3429–3434. https://doi.org/10.1128/AEM.72.5.3429-3434.2006

Martins, M. R., Jantalia, C. P., Reis, V. M., Döwich, I., Polidoro, J. C., Alves, B. J. R., Boddey, R. M., & Urquiaga, S. (2018). Impact of plant growth-promoting bacteria on grain yield, protein content, and urea-15 N recovery by maize in a cerrado oxisol. *Plant and Soil, 422*(1–2), 239–250. https://doi.org/10.1007/s11104-017-3193-1

Maruri-López, I., Aviles-Baltazar, N. Y., Buchala, A., & Serrano, M. (2019). Intra and extracellular journey of the phytohormone salicylic acid. *Frontiers in Plant Science, 10*(April), 1–11. https://doi.org/10.3389/fpls.2019.00423

Maske, B. L., de Melo Pereira, G. V., da S. Vale, A., de Carvalho Neto, D. P., Karp, S. G., Viesser, J. A., De Dea Lindner, J., Pagnoncelli, M. G., Soccol, V. T., & Soccol, C. R. (2021). A review on enzyme-producing lactobacilli associated with the human digestive process: From metabolism to application. *Enzyme and Microbial Technology, 149*(May). https://doi.org/10.1016/j.enzmictec.2021.109836

Masschelein, J., Jenner, M., & Challis, G. L. (2017). Natural product reports antibiotics from gram-negative bacteria: A comprehensive overview and selected. *Natural Product Reports, 00*, 1–72. https://doi.org/10.1039/C7NP00010C

Massart, S., Perazzolli, M., Höfte, M., Pertot, I., & Jijakli, M. H. (2015). Impact of the omic technologies for understanding the modes of action of biological control agents against plant pathogens. *BioControl, 60*(6), 725–746. https://doi.org/10.1007/s10526-015-9686-z

Matei, A., & Doehlemann, G. (2016). Cell biology of corn smut disease — ustilago maydis as a model for biotrophic interactions. *Current Opinion in Microbiology, 34*, 60–66. https://doi.org/10.1016/j.mib.2016.07.020

Matson, P. A. (2012). Seeds of sustainability. In P. A. Matson (Ed.), *Seeds of sustainability* (1st ed.). Island Press.

Matsumoto, H., Fan, X., Wang, Y., Kusstatscher, P., Duan, J., Wu, S., Chen, S., Qiao, K., Wang, Y., Ma, B., Zhu, G., Hashidoko, Y., Berg, G., Cernava, T., & Wang, M. (2021). Bacterial seed endophyte shapes disease resistance in rice. *Nature Plants, 7*(1), 60–72. https://doi.org/10.1038/s41477-020-00826-5

Mauch-Mani, B., Baccelli, I., Luna, E., & Flors, V. (2017). Defense priming: An adaptive part of induced resistance. *Annual Review of Plant Biology, 68*(January), 485–512. https://doi.org/10.1146/annurev-arplant-042916-041132

McNeely, D., Chanyi, R. M., Dooley, J. S., Moore, J. E., Koval, S. F., & Author. (2016). Biocontrol of Burkholderia cepacia complex bacteria and bacterial phytopathogens by bdellovibrio bacteriovorus. *Canadian Journal of Microbiology, 63*(4), 350–358.

Meena, B. (2014). Biological control of pest and diseases using fluorescent pseudomonads. In K. Sahayaraj (Ed.), *Basic and applied aspects of biopesticides* (pp. 17–29). Springer.

Meier-Kolthoff, J. P., & Göker, M. (2019). TYGS is an automated high-throughput platform for state-of-the-art genome-based taxonomy. *Nature Communications, 10*(1). https://doi.org/10.1038/s41467-019-10210-3

Meier-Kolthoff, J. P., Auch, A. F., Klenk, H. P., & Göker, M. (2013). Genome sequence-based species delimitation with confidence intervals and improved distance functions. *BMC Bioinformatics, 14*. https://doi.org/10.1186/1471-2105-14-60

Meier-Kolthoff, J. P., Hahnke, R. L., Petersen, J., Scheuner, C., Michael, V., Fiebig, A., Rohde, C., Rohde, M., Fartmann, B., Goodwin, L. A., Chertkov, O., Reddy, T., Pati, A., Ivanova, N. N., Markowitz, V., Kyrpides, N. C., Woyke, T., Göker, M., & Klenk, H.-P. (2014). Complete genome sequence of DSM 30083T, the proposal for delineating subspecies in microbial taxonomy. *Standards in Genomic Sciences, 1–19.*

Mendis, H. C., Thomas, V. P., Schwientek, P., Salamzade, R., Chien, J.-T., Waidyarathne, P., Kloepper, J., & De La Fuente, L. (2018). Strain-specific quantification of root colonization by plant growth promoting rhizobacteria *Bacillus firmus* I-1582 and *Bacillus amyloliquefaciens* QST713 in non-sterile soil and field conditions. *PLoS One, 13*(2), e0193119. https://doi.org/10.1371/journal.pone.0193119

Messing, S. A. J., Mario Amzel, L., Gabelli, S. B., Echeverria, I., Vogel, J. T., Guan, J. C., Tan, B. C., Klee, H. J., & McCarty, D. R. (2010). Structural insights into maize viviparous14, a key enzyme in the biosynthesis of the phytohormone abscisic acid. *Plant Cell, 22*(9), 2970–2980. https://doi.org/10.1105/tpc.110.074815

Micheletti, M., & Lye, G. J. (2006). Microscale bioprocess optimisation. *Current Opinion in Biotechnology, 17*(6), 611–618. https://doi.org/10.1016/j.copbio.2006.10.006

Millan, A. F. S., Gamir, J., Farran, I., Larraya, L., & Veramendi, J. (2022). Identification of new antifungal metabolites produced by the yeast *Metschnikowia pulcherrima* involved in the biocontrol of postharvest plant pathogenic fungi. *Postharvest Biology and Technology, 192*(March), 111995. https://doi.org/10.1016/j.postharvbio.2022.111995

Mishra, N., & Sundari, K. (2015). Native PGPM consortium: A beneficial solution to support plant growth in the presence of phytopathogens and residual organophosphate pesticides. *Journal of Bioprocessing and Biotechniques, 05*(02). https://doi.org/10.4172/2155-9821.1000202

Mitra, D., Anđelković, S., Panneerselvam, P., Senapati, A., Vasić, T., Ganeshamurthy, A. N., Chauhan, M., Uniyal, N., Mahakur, B., & Radha, T. K. (2020). Phosphate-Solubilizing microbes and biocontrol agent for plant nutrition and protection: Current perspective. *Communications in Soil Science and Plant Analysis, 51*(5), 645–657. https://doi.org/10.1080/00103624.2020.1729379

Mitter, B., Brader, G., Pfaffenbichler, N., & Sessitsch, A. (2019). Next generation microbiome applications for crop production — limitations and the need of knowledge-based solutions. *Current Opinion in Microbiology, 49*, 59–65. https://doi.org/10.1016/j.mib.2019.10.006

Mizrahi-Man, O., Davenport, E. R., & Gilad, Y. (2013). Taxonomic classification of bacterial 16S rRNA genes using short sequencing reads: Evaluation of effective study designs. *PLoS ONE, 8*(1), 18–23. https://doi.org/10.1371/journal.pone.0053608

Moeinzadeh, A., Sharif-Zadeh, F., Ahmadzadeh, M., & Tajabadi, F. H. (2010). Biopriming of sunflower (*Helianthus annuus* L.) seed with pseudomonas fluorescens for improvement of seed invigoration and seedling growth. *Australian Journal of Crop Science, 4*(7), 564–570.

Montaño López, J., Duran, L., & Avalos, J. L. (2022). Physiological limitations and opportunities in microbial metabolic engineering. *Nature Reviews Microbiology, 20*(1), 35–48. https://doi.org/10.1038/s41579-021-00600-0

Monteiro, S. M. S., Clemente, J. J., Carrondo, M. J. T., & Cunha, A. E. (2014). Enhanced spore production of *Bacillus subtilis* grown in a chemically defined medium. *Advances in Microbiology, 04*(08), 444–454. https://doi.org/10.4236/aim.2014.48049

Moore, J. W., Loake, G. J., & Spoel, S. H. (2011). Transcription dynamics in plant immunity. *Plant Cell, 23*(8), 2809–2820. https://doi.org/10.1105/tpc.111.087346

Moradi, S., Khoshru, B., Mitra, D., Mahakur, B., Das Mohapatra, P. K., Asgari Lajayer, B., & Ghorbanpour, M. (2021). Transcriptomics analyses and the relationship between plant and plant growth-promoting rhizobacteria (PGPR). In R. N. Pudake, B. B. Sahu, M. Kumari, & A. K. Sharma (Eds.), *Omics science for rhizosphere biology* (pp. 89–111). Springer Nature. https://doi.org/10.1007/978-981-16-0889-6_6

Morales, P., Valenzuela, V., Ortega, M., Martínez, A., Félix, C., Chávez, R., Parra, F., & Santos, S. (2021). Taxonomía bacteriana basada en índices relacionados al genoma completo. *La Sociedad Académica, 58*(December). https://www.itson.mx/publicaciones/sociedad-academica/Documents/LSA58_compressed.pdf#page=41.

Mordor Intelligence Research, Advisory. (June 2023). Biofertilizers Market Size & Share Analysis - Growth Trends & Forecasts (2023 - 2028. *Mordor Intelligence.* https://www.mordorintelligence.com/industry-reports/global-biofertilizers-market-industry.

Moreno-Reséndez, A., García-Mendoza, V., Reyes-Carrillo, J. L., Vásquez-Arroyo, J., & Cano-Ríos, P. (2018). Rizobacterias promotoras del crecimiento vegetal: Una alternativa de biofertilización para la agricultura sustentable. *Revista Colombiana de Biotecnología, 20*(1), 68–83. https://doi.org/10.15446/rev.colomb.biote.v20n1.73707

Moreno-Ruiz, D., Lichius, A., Turrà, D., Di Pietro, A., & Zeilinger, S. (2020). Chemotropism assays for plant symbiosis and mycoparasitism related compound screening in trichoderma atroviride. *Frontiers in Microbiology, 11*(November), 1–17. https://doi.org/10.3389/fmicb.2020.601251

Moriuchi, R., Dohra, H., Kanesaki, Y., & Ogawa, N. (2019). Complete genome sequence of 3-chlorobenzoate-degrading bacterium cupriavidus necator NH9 and reclassification of the strains of the genera cupriavidus and ralstonia based on phylogenetic and whole-genome sequence analyses. *Frontiers in Microbiology, 10*(February), 1–21. https://doi.org/10.3389/fmicb.2019.00133

Moriya, Y., Itoh, M., Okuda, S., Yoshizawa, A. C., & Kanehisa, M. (2007). Kaas: An automatic genome annotation and pathway reconstruction server. *Nucleic Acids Research, 35*(2), 182–185. https://doi.org/10.1093/nar/gkm321

Mosquera, G., Giraldo, M. C., Khang, C. H., Coughlan, S., & Valent, B. (2009). Interaction transcriptome analysis identifies magnaporthe oryzae BAS1-4 as biotrophy-associated secreted proteins in rice blast disease. *Plant Cell, 21*(4), 1273–1290. https://doi.org/10.1105/tpc.107.055228

Mourya, S., & Jauhri, K. S. (2002). LacZ tagging of phosphate solubilizing *Pseudomonas striata* for rhizosphere colonization. *Indian Journal of Biotechnology, 1*(3), 275–279.

Muhammad, T., Zhang, F., Zhang, Y., & Liang, Y. (2019). RNA interference: A natural immune system of plants to counteract biotic stressors. *Cells, 8*(1), 38. https://doi.org/10.3390/cells8010038

Murphy, J. F., Reddy, M. S., Ryu, C.-M., Kloepper, J. W., & Li, R. (2003). Rhizobacteria-mediated growth promotion of tomato leads to protection against cucumber mosaic virus. *Phytopathology, 93*(10), 1301–1307. https://doi.org/10.1094/PHYTO.2003.93.10.1301

Nadeem, S. M., Ahmad, M., Zahir, Z. A., Javaid, A., & Ashraf, M. (2014). The role of mycorrhizae and plant growth promoting rhizobacteria (PGPR) in improving crop productivity under stressful environments. *Biotechnology Advances, 32*(2), 429–448. https://doi.org/10.1016/j.biotechadv.2013.12.005

Namvar, A., Haq, I., Shields, M., Amoako, K. K., & Warriner, K. (2013). Extraction of *Bacillus endospores* from water, apple juice concentrate, raw milk and lettuce rinse solutions using tangential flow filtration. *Food Control, 32*(2), 632–637. https://doi.org/10.1016/j.foodcont.2013.01.033

Navarro, E., Serrano-Heras, G., Castaño, M. J., & Solera, J. (2015). Real-time PCR detection chemistry. *Clinica Chimica Acta, 439*, 231–250. https://doi.org/10.1016/j.cca.2014.10.017

Narayanasamy, P., & Agents, B. C. (2013). Mechanisms of action of bacterial biological control agents. In P. Narayanasamy (Ed.), *Biological management of diseases of crops. Volume 1: Characteristics of biological control agents* (Vol. 1). Springer.

Navarro-Muñoz, J. C., Selem-Mojica, N., Mullowney, M. W., Kautsar, S. A., Tryon, J. H., Parkinson, E. I., De Los Santos, E. L. C., Yeong, M., Cruz-Morales, P., Abubucker, S., Roeters, A., Lokhorst, W., Fernandez-Guerra, A., Cappelini, L. T. D., Goering, A. W., Thomson, R. J., Metcalf, W. W., Kelleher, N. L., Barona-Gomez, F., & Medema, M. H. (2020). A computational framework to explore large-scale biosynthetic diversity. *Nature Chemical Biology, 16*(1), 60–68. https://doi.org/10.1038/s41589-019-0400-9

Neemisha. (2020). Role of soil organisms in maintaining soil health, ecosystem functioning, and sustaining agricultural production. In B. Giri, & A. Varma (Eds.), *Soil Health. Soil Biology* (Vol 59). Cham: Springer. https://doi.org/10.1007/978-3-030-44364-1_17

Negi, Y. K., Garg, S. K., & Kumar, J. (2005). Cold-tolerant fluorescent pseudomonas isolates from Garhwal Himalayas as potential plant growth promoting and biocontrol agents in pea Cold-tolerant fluorescent pseudomonas isolates from Garhwal Himalayas as potential plant growth promoting and biocontrol. *Current Science, 89*(12), 2151–2156.

Negus, D., Moore, C., Baker, M., Raghunathan, D., Tyson, J., & Sockett, R. E. (2017). Predator versus pathogen: How does predatory bdellovibrio bacteriovorus interface with the challenges of killing gram-negative pathogens in a host setting? *Annual Review of Microbiology, 71*, 441–457. https://doi.org/10.1146/annurev-micro-090816-093618

Nejat, N., Rookes, J., Mantri, N. L., & Cahill, D. M. (2017). Plant–pathogen interactions: Toward development of next-generation disease-resistant plants. *Critical Reviews in Biotechnology, 37*(2), 229–237. https://doi.org/10.3109/07388551.2015.1134437

Nelson, R., Wiesner-Hanks, T., Wisser, R., & Balint-Kurti, P. (2018). Navigating complexity to breed disease-resistant crops. *Nature Reviews Genetics, 19*(1), 21–33. https://doi.org/10.1038/nrg.2017.82

Newman, M.-A., Sundelin, T., Nielsen, J. T., & Erbs, G. (2013). MAMP (microbe-associated molecular pattern) triggered immunity in plants. *Frontiers in Plant Science, 4*(May), 1–14. https://doi.org/10.3389/fpls.2013.00139

Ney, L., Franklin, D., Mahmud, K., Cabrera, M., Hancock, D., Habteselassie, M., Newcomer, Q., & Dahal, S. (2020). Impact of inoculation with local effective microorganisms on soil nitrogen cycling and legume productivity using composted broiler litter. *Applied Soil Ecology, 154*(February), 103567. https://doi.org/10.1016/j.apsoil.2020.103567

Nguyen, T. H., Phan, T. C., Choudhury, A. T. M. A., Rose, M. T., Deaker, R. J., & Kennedy, I. R. (2017). BioGro: A plant growth-promoting biofertilizer validated by 15 Years' research from laboratory selection to rice farmer's fields of the mekong delta. In J. Singh, & G. Seneviratne (Eds.), *Agro-Environmental sustainability* (pp. 237–257). Springer. https://doi.org/10.1007/978-3-319-49724-2

Niderman, T., Genetet', L., Bruyère, T., Gees, R., Stintzi, A., Legrand, M., Fritig, B., & Mosinger, E. (1995). Pathogenesis-Related PR-1 proteins are antifungal lsolation and characterization of three 14-kilodalton proteins of tomato and of a basic PR-1 of tobacco with inhibitory activity against phyfophfhora infestans. In *Plant physiol* (Vol. 108). https://academic.oup.com/plphys/article/108/1/17/6069614.

Niu, B., Paulson, J. N., Zheng, X., & Kolter, R. (2017). Simplified and representative bacterial community of maize roots. *Proceedings of the National Academy of Sciences of the United States of America, 114*(12), E2450–E2459. https://doi.org/10.1073/pnas.1616148114

Niu, B., Wang, W., Yuan, Z., Sederoff, R. R., Sederoff, H., Chiang, V. L., & Borriss, R. (2020). Microbial interactions within multiple-strain biological control agents impact soil-borne plant disease. *Frontiers in Microbiology, 11*(October), 1–16. https://doi.org/10.3389/fmicb.2020.585404

Odoh, C. K., Sam, K., Zabbey, N., Eze, C. N., Nwankwegu, A. S., Laku, C., Dumpe, B. B., & Abstract. (2020). Microbial consortium as biofertilizers for crops growing under the extreme habitats. In A. N. Yadav, J. Singh, A. A. Rastegari, & N. Yadav (Eds.), *Plant microbiomes for sustainable agriculture, sustainable development and biodiversity* (pp. 381–424). Springer. https://doi.org/10.1007/978-3-030-38453-1

Olanrewaju, O. S., Glick, B. R., & Babalola, O. O. (2017). Mechanisms of action of plant growth promoting bacteria. *World Journal of Microbiology and Biotechnology, 33*(11), 1–16. https://doi.org/10.1007/s11274-017-2364-9

OMPI. (2017). *Tratado de Budapest sobre el Reconocimiento Internacional del Depósito de Microorganismos a los fines del Procedimiento en materia de Patentes*. https://www.wipo.int/treaties/es/registration/budapest/index.html.

Ooijkaas, L. P., Wilkinson, E. C., Tramper, J., & Buitelaar, R. M. (1999). Medium optimization for spore production of *Coniothyrium minitans* using statistically-based experimental designs. *Biotechnology and Bioengineering, 64*(1), 92–100. https://doi.org/10.1002/(SICI)1097-0290(19990705)64:1<92::AID-BIT10>3.0.CO;2-8

Orozco-Alvarez, C., García-Salas, S., & Moreno-Rivera, L. (2013). *Tangential flow filtration of spores suspension.*

Oteino, N., Lally, R. D., Kiwanuka, S., Lloyd, A., Ryan, D., Germaine, K. J., & Dowling, D. N. (2015). Plant growth promotion induced by phosphate solubilizing endophytic Pseudomonas isolates. *Frontiers in Microbiology, 6*(July), 1–9. https://doi.org/10.3389/fmicb.2015.00745

Pandey, A., Sharma, E., & Palni, L. M. S. (1998). Influence of bacterial inoculation on maize in upland farming systems of the Sikkim Himalaya. *Soil Biology and Biochemistry, 30*(3), 379–384.

Pang, Q., Zhang, T., Wang, Y., Kong, W., Guan, Q., Yan, X., & Chen, S. (2018). Metabolomics of early stage plant cell–microbe interaction using stable isotope labeling. *Frontiers in Plant Science, 9*(June), 1–11. https://doi.org/10.3389/fpls.2018.00760

Pang, Z., Zhou, G., Ewald, J., Chang, L., Hacariz, O., Basu, N., & Xia, J. (2022). Using MetaboAnalyst 5.0 for LC–HRMS spectra processing, multi-omics integration and covariate adjustment of global metabolomics data. *Nature Protocols, 17*(8), 1735–1761. https://doi.org/10.1038/s41596-022-00710-w

Parasuraman, P., Pattnaik, S., & Busi, S. (2019). Phyllosphere microbiome: Functional importance in sustainable agriculture. In J. S. Singh, & D. P. Singh (Eds.), *New and future developments in microbial biotechnology and bioengineering: Microbial biotechnology in agro-environmental sustainability* (pp. 135–148). Elsevier B.V. https://doi.org/10.1016/B978-0-444-64191-5.00010-9

Parnell, J., Berka, R., Young, H., Sturino, J., Kang, Y., Barnhart, D., & Dileo, M. (2016). From the lab to the farm: An industrial perspective of plant beneficial microorganisms. *Frontiers in Plant Science, 7*, 1110. https://doi.org/10.3389/fpls.2016.01110

Parra-Cota, F. I., Coronel-Acosta, C.-B., Amézquita-Avilés, C. F., De los Santos-Villalobos, S., & Escalante-Martínez, D. I. (2018). Diversidad metabólica de microorganismos edáficos asociados al cultivo de maíz en el Valle del Yaqui, Sonora. *Revista Mexicana de Ciencias Agrícolas, 9*(2), 431–442. https://doi.org/10.29312/remexca.v9i2.1083

Pascale, A., Proietti, S., Pantelides, I. S., & Stringlis, I. A. (2020). Modulation of the root microbiome by plant molecules: The basis for targeted disease suppression and plant growth promotion. *Frontiers in Plant Science, 10*(January), 1–23. https://doi.org/10.3389/fpls.2019.01741

Passera, A., Vacchini, V., Cocetta, G., Shahzad, G. I. R., Arpanahi, A. A., Casati, P., Ferrante, A., & Piazza, L. (2020). Towards Nutrition-Sensitive Agriculture: An evaluation of biocontrol effects, nutritional value, and ecological impact of bacterial inoculants. *Science of the Total Environment, 724*, 138127. https://doi.org/10.1016/j.scitotenv.2020.138127

Paterson, J., Jahanshah, G., Li, Y., Wang, Q., Mehnaz, S., & Gross, H. (2017). The contribution of genome mining strategies to the understanding of active principles of PGPR strains. *FEMS Microbiology Ecology, 93*(3), 1–12. https://doi.org/10.1093/femsec/fiw249

Patil, H. J., & Solanki, M. K. (2016). Microbial inoculant: Modern era of fertilizers and pesticides. In D. Singh, H. Singh, & R. Prabha (Eds.), *Microbial inoculants in sustainable agricultural productivity* (pp. 319–343). Springer. https://doi.org/10.1007/978-81-322-2647-5

Patz, S., Gautam, A., Becker, M., Ruppel, S., Rodríguez-Palenzuela, P., & Huson, D. (2021). PLaBAse: A comprehensive web resource for analyzing the plant growth-promoting potential of plant-associated bacteria. *BioRxiv, 3*, 2021.12.13.472471 https://www.biorxiv.org/content/10.1101/2021.12.13.472471v1%0Ahttps://www.biorxiv.org/content/10.1101/2021.12.13.472471v1.abstract.

Paul, S., & Das, S. (2021). Natural insecticidal proteins, the promising bio-control compounds for future crop protection. *Nucleus (India), 64*(1), 7–20. https://doi.org/10.1007/s13237-020-00316-1

Peer, R.van (1991). Induced resistance and phytoalexin accumulation in biological control of fusarium wilt of carnation by pseudomonas sp. strain WCS417r. *Phytopathology, 81*(7), 728. https://doi.org/10.1094/phyto-81-728

Peiffer, J. A., Spor, A., Koren, O., Jin, Z., Tringe, S. G., Dangl, J. L., Buckler, E. S., & Ley, R. E. (2013). Diversity and heritability of the maize rhizosphere microbiome under field conditions. *Proceedings of the National Academy of Sciences of the United States of America, 110*(16), 6548–6553. https://doi.org/10.1073/pnas.1302837110

Peláez-Álvarez, A., Santos-Villalobos, S. de los, Yépez, E. A., Parra-Cota, F. I., & Reyes-Rodríguez, R. T. (2016). Efecto sinérgico de Trichoderma asperellum T8A y captan 50® contra Colletotrichum gloeosporioides (Penz.) TT - synergistic effect of Trichoderma asperelleum T8A and captan 50® against colletotrichum gloeosporioides (Penz.). *Revista Mexicana de Ciencias Agrícolas, 7*(6), 1401–1412. http://www.scielo.org.mx/scielo.php?script=sci_arttext&pid=S2007-09342016000601401&lang=pt.%0Ahttp://www.scielo.org.mx/pdf/remexca/v7n6/2007-0934-remexca-7-06-1401.pdf.

Peng, A., Chen, S., Lei, T., Xu, L., He, Y., Wu, L., Yao, L., & Zou, X. (2017). Engineering canker-resistant plants through CRISPR/Cas9-targeted editing of the susceptibility gene CsLOB1 promoter in citrus. *Plant Biotechnology Journal, 15*(12), 1509–1519. https://doi.org/10.1111/pbi.12733

Peng, J., Wu, D., Liang, Y., Li, L., & Guo, Y. (2019). Disruption of acdS gene reduces plant growth promotion activity and maize saline stress resistance by *Rahnella aquatilis* HX2. *Journal of Basic Microbiology, 59*(4), 402–411. https://doi.org/10.1002/jobm.201800510

Perazzolli, M., Moretto, M., Fontana, P., Ferrarini, A., Velasco, R., Moser, C., Delledonne, M., & Pertot, I. (2012). Downy mildew resistance induced by *Trichoderma harzianum* T39 in susceptible grapevines partially mimics transcriptional changes of resistant genotypes. *BMC Genomics, 13*(1), 1–19. https://doi.org/10.1186/1471-2164-13-660

Pérez Peñaranda, M. C., Oramas García, J., Sotolongo Valdés, E. A., Galuzzo, A. M., Román Tabio, Y., & González Soto, A. (2019). Optimización del medio de cultivo y las condiciones de fermentación para la producción de un biofertilizante a base de pseudomonas fluorescens. *Biotecnología Vegetal, 19*(2), 127–138. http://scielo.sld.cu/scielo.php?script=sci_arttext&pid=S2074-86472019000200127.

Pham, J. V., Yilma, M. A., Feliz, A., Majid, M. T., Maffetone, N., Walker, J. R., Kim, E., Cho, H. J., Reynolds, J. M., Song, M. C., Park, S. R., & Yoon, Y. J. (2019). A review of the microbial production of bioactive natural products and biologics. *Frontiers in Microbiology, 10*(June), 1–27. https://doi.org/10.3389/fmicb.2019.01404

Philippot, L., Raaijmakers, J. M., Lemanceau, P., & Van Der Putten, W. H. (2013). Going back to the roots: The microbial ecology of the rhizosphere. *Nature Reviews Microbiology, 11*(11), 789–799. https://doi.org/10.1038/nrmicro3109

Phukan, U. J., Jeena, G. S., & Shukla, R. K. (2016). WRKY transcription factors: Molecular regulation and stress responses in plants. *Journal of Integral Plant Biology, 7*(June), 1–14. https://doi.org/10.3389/fpls.2016.00760

Pieterse, C. M. J., & Van Loon, L. C. (1999). Salicylic acid-independent plant defence pathways. *Trends in Plant Science, 4*(2), 52–58. https://doi.org/10.1016/S1360-1385(98)01364-8

Pieterse, C. M. J., Wees, S. C. M. Van, Hoffland, E., Pelt, J. A. Van, & Loon, L. C. Van (1996). American society of plant biologists (ASPB) acid accumulation and pathogenesis-related gene expression. *The Plant Cell, 8*(8), 1225–1237.

Pieterse, C. M. J., Van Wees, S. C. M., Van Pelt, J. A., Knoester, M., Laan, R., Gerrits, H., Weisbeek, P. J., & Van Loon, L. C. (1998). A novel signaling pathway controlling induced systemic resistance in arabidopsis. *Plant Cell, 10*(9), 1571–1580. https://doi.org/10.1105/tpc.10.9.1571

Pieterse, C. M. J., Van Der Does, D., Zamioudis, C., Leon-Reyes, A., & Van Wees, S. C. M. (2012). Hormonal modulation of plant immunity. *Annual Review of Cell and Developmental Biology, 28*(May), 489–521. https://doi.org/10.1146/annurev-cellbio-092910-154055

Pieterse, C. M. J., Zamioudis, C., Berendsen, R. L., Weller, D. M., Van Wees, S. C. M., & Bakker, P. A. H. M. (2014). Induced systemic resistance by beneficial microbes. *Annual Review of Phytopathology, 52*(1), 347–375. https://doi.org/10.1146/annurev-phyto-082712-102340

Pixley, K. V., Falck-Zepeda, J. B., Giller, K. E., Glenna, L. L., Gould, F., Mallory-Smith, C. A., Stelly, D. M., & Stewart, C. N. (2019). Genome editing, gene drives, and synthetic biology: Will they contribute to disease-resistant crops, and who will benefit? *Annual Review of Phytopathology, 57*, 165–188. https://doi.org/10.1146/annurev-phyto-080417-045954

Posada-Uribe, L. F., Romero-Tabarez, M., & Villegas-Escobar, V. (2015). Effect of medium components and culture conditions in *Bacillus subtilis* EA-CB0575 spore production. *Bioprocess and Biosystems Engineering, 38*(10). https://doi.org/10.1007/s00449-015-1428-1

Prakash, O., Verma, M., Sharma, P., Kumar, M., Kumari, K., Singh, A., Kumari, H., Jit, S., Gupta, S. K., Khanna, M., & Lal, R. (2007). Polyphasic approach of bacterial classification - an overview of recent advances. *Indian Journal of Microbiology, 47*(2), 98–108. https://doi.org/10.1007/s12088-007-0022-x

Pretty, J., & Bharucha, Z. P. (1979). Sustainable intensification in agricultural systems. *Annals of Botany, 114*(8), 1571–1596. https://doi.org/10.1093/aob/mcu205

Preininger, C., Sauer, U., Bejarano, A., & Berninger, T. (2018). Concepts and applications of foliar spray for microbial inoculants. *Applied Microbiology and Biotechnology, 102*, 7265–7282.

Prins, T. W., Tudzynski, P., von Tiedemann, A., Tudzynski, B., Ten Have, A., Hansen, M. E., Tenberge, K., & van Kan, J. A. L. (2000). Infection strategies of botrytis cinerea and related necrotrophic pathogens. *Fungal Pathology*, 33–64. https://doi.org/10.1007/978-94-015-9546-9_2

Prjibelski, A., Antipov, D., Meleshko, D., Lapidus, A., & Korobeynikov, A. (2020). Using SPAdes de novo assembler. *Current Protocols in Bioinformatics, 70*(1), 1–29. https://doi.org/10.1002/cpbi.102

Proctor, R. H., Hohn, T. M., & McCormick, S. P. (1995). Reduced virulence of Gibberella zeae caused by disruption of a trichothecene toxin biosynthetic gene. In *Molecular plant-microbe interactions* (Vol. 8, pp. 593–601). https://doi.org/10.1094/MPMI-8-0593

Pruitt, R. N., Gust, A. A., & Nürnberger, T. (2021). Plant immunity unified. *Nature Plants, 7*(4), 382–383. https://doi.org/10.1038/s41477-021-00903-3

Qiu, M., Xu, Z., Li, X., Li, Q., Zhang, N., Shen, Q., & Zhang, R. (2014). Comparative proteomics analysis of Bacillus amyloliquefaciens SQR9 revealed the key proteins involved in in situ root colonization. *Journal of Proteome Research, 13*(12). https://doi.org/10.1021/pr500565m

Qiu, Z., Egidi, E., Liu, H., Kaur, S., & Singh, B. K. (2019). New frontiers in agriculture productivity: Optimised microbial inoculants and in situ microbiome engineering. *Biotechnology Advances, 37*(6), 107371. https://doi.org/10.1016/j.biotechadv.2019.03.010

Qu, Q., Zhang, Z., Peijnenburg, W. J. G. M., Liu, W., Lu, T., Hu, B., Chen, J., Chen, J., Lin, Z., & Qian, H. (2020). Rhizosphere microbiome assembly and its impact on plant growth. *Journal of Agricultural and Food Chemistry, 68*(18), 5024–5038. https://doi.org/10.1021/acs.jafc.0c00073

Raaijmakers, J. M., & Mazzola, M. (2012). *Diversity and natural functions of antibiotics produced by beneficial and plant pathogenic bacteria*. https://doi.org/10.1146/annurev-phyto-081211-172908. *May*, 1–22.

Radzki, W., Gutierrez Mañero, F. J., Algar, E., Lucas García, J. A., García-Villaraco, A., & Ramos Solano, B. (2013). Bacterial siderophores efficiently provide iron to iron-starved tomato plants in hydroponics culture. *Antonie Van Leeuwenhoek, International Journal of General and Molecular Microbiology, 104*(3), 321–330. https://doi.org/10.1007/s10482-013-9954-9

Rahman, A., & Pachter, L. (2013). Cgal: Computing genome assembly likelihoods. *Genome Biology, 14*(1), 1–10. https://doi.org/10.1186/gb-2013-14-1-r8

Rajkumar, M., Ae, N., Prasad, M. N. V., & Freitas, H. (2010). Potential of siderophore-producing bacteria for improving heavy metal phytoextraction. *Trends in Biotechnology, 28*(3), 142–149. https://doi.org/10.1016/j.tibtech.2009.12.002

Ramesh, K., Nazeer, A., Desh, B. S., Om, C., Sharma, Shiv, L., & Mohammad, M. S. (2013). Enhancing blooming period and propagation coefficient of tulip (*Tulipa gesneriana* L.) using growth regulators. *African Journal of Biotechnology, 12*(2), 168–174. https://doi.org/10.5897/ajb12.2713

Ramlucken, U., Ramchuran, S. O., Moonsamy, G., Jansen van Rensburg, C., Thantsha, M. S., & Lalloo, R. (2021). Production and stability of a multi-strain Bacillus based probiotic product for commercial use in poultry. *Biotechnology Reports, 29*, e00575. https://doi.org/10.1016/j.btre.2020.e00575

Rasul, M., Yasmin, S., Yahya, M., Breitkreuz, C., Tarkka, M., & Reitz, T. (2021). The wheat growth-promoting traits of *Ochrobactrum* and *Pantoea* species, responsible for solubilization of different P sources, are ensured by genes encoding enzymes of multiple P-releasing pathways. *Microbiological Research, 246*(December), 126703. https://doi.org/10.1016/j.micres.2021.126703

Ravikumar, S., Kathiresan, K., Liakath Alikhan, S., Prakash Williams, G., & Anitha Anandha Gracelin, N. (2007). Growth of *Avicennia marina* and *Ceriops decandra* seedlings inoculated with halophilic azotobacters. *Journal of Environmental Biology, 28*(3), 601–603.

Reetha, D., Kumaresan, G., & John Milton, D. (2014). Studies to improve the shelf life of *Azospirillum lipoferum* immobilized in alginate beads. *International Journal of Recent Scientific Research, 5*(12), 2178–2182.

Reichheld, J. P., Vernoux, T., Lardon, F., Van Montagu, M., & Inzé, D. (1999). Specific checkpoints regulate plant cell cycle progression in response to oxidative stress. *Plant Journal, 17*(6), 647–656. https://doi.org/10.1046/j.1365-313X.1999.00413.x

Ren, B., Liang, Y., Deng, Y., Chen, Q., Zhang, J., Yang, X., & Zuo, J. (2009). Genome-wide comparative analysis of type-A Arabidopsis response regulator genes by overexpression studies reveals their diverse roles and regulatory mechanisms in cytokinin signaling. *Cell Research, 19*(10), 1178–1190. https://doi.org/10.1038/cr.2009.88

Rep, M., Van Der Does, H. C., Meijer, M., Van Wijk, R., Houterman, P. M., Dekker, H. L., De Koster, C. G., & Cornelissen, B. J. C. (2004). A small, cysteine-rich protein secreted by fusarium oxysporum during colonization of xylem vessels is required for I-3-mediated resistance in tomato. *Molecular Microbiology, 53*(5), 1373–1383. https://doi.org/10.1111/j.1365-2958.2004.04177.x

Richter, M., & Rosselló-Móra, R. (2009). Shifting the genomic gold standard for the prokaryotic species definition. *Proceedings of the National Academy of Sciences of the United States of America, 106*(45), 19126–19131. https://doi.org/10.1073/pnas.0906412106

Richter, M., Rosselló-Móra, R., Oliver Glöckner, F., & Peplies, J. (2016). JSpeciesWS: A web server for prokaryotic species circumscription based on pairwise genome comparison. *Bioinformatics, 32*(6), 929–931. https://doi.org/10.1093/bioinformatics/btv681

Rilling, J., Acuña, J. A., Nannipieri, P., Cassan, F., Maruyama, F., & Jorquera, M. (2019). Current opinion and perspectives on the methods for tracking and monitoring plant growth–promoting bacteria. *Soil Biology and Biochemistry, 130*, 205–219. https://doi.org/10.1016/j.soilbio.2018.12.012

Rinu, K., & Pandey, A. (2009). Bacillus subtilis NRRL B-30408 inoculation enhances the symbiotic efficiency of Lens esculenta Moench at a Himalayan location. *Journal of Plant Nutrition and Soil Science, 172*, 134–139. https://doi.org/10.1002/jpln.200800153

Rissman, A. I., Mau, B., Biehl, B. S., Darling, A. E., Glasner, J. D., & Perna, N. T. (2009). Reordering contigs of draft genomes using the Mauve Aligner. *Bioinformatics, 25*(16), 2071–2073. https://doi.org/10.1093/bioinformatics/btp356

Ritchie, H., & Roser, M. (2019). Crop Yields. *Our World in Data*. https://ourworldindata.org/crop-yields.

Robles-Montoya, R. I., Chaparro-Encinas, L. A., Parra-Cota, F. I., & de los Santos-Villalobos, S. (2020). Improving biometric traits of wheat seedlings with the inoculation of a consortium native of *Bacillus. Revista Mexicana Ciencias Agrícolas, 11*(1), 229–235.

Robles-Montoya, R. I., Valenzuela-Ruiz, V., Parra-Cota, F. I., Santoyo, G., & de los Santos-Villalobos, S. (2020). Description of a polyphasic taxonomic approach for plant growth-promoting rhizobacteria (PGPR). In J. S. Singh, & S. R. Vimal (Eds.), *Microbial services in restoration ecology* (pp. 259–269). Elsevier.

Roca, A., Pizarro-Tobías, P., Udaondo, Z., Fernández, M., Matilla, M. A., Molina-Henares, M. A., Molina, L., Segura, A., Duque, E., & Ramos, J. L. (2013). Analysis of the plant growth-promoting properties encoded by the genome of the rhizobacterium Pseudomonas putida BIRD-1. *Environmental Microbiology, 15*(3), 780–794. https://doi.org/10.1111/1462-2920.12037

Rocha, I., Ma, Y., Souza-Alonso, P., Vosátka, M., Freitas, H., & Oliveira, R. S. (2019). Seed coating: A tool for delivering beneficial microbes to agricultural crops. *Frontiers in Plant Science, 10*(1357). https://doi.org/10.3389/fpls.2019.01357

Rodriguez, P. A., Rothballer, M., Chowdhury, S. P., Nussbaumer, T., Gutjahr, C., & Falter-Braun, P. (2019). Systems biology of plant-microbiome interactions. *Molecular Plant, 12*(6), 804–821. https://doi.org/10.1016/j.molp.2019.05.006

Rodriguez-Moreno, L., Ebert, M. K., Bolton, M. D., & Thomma, B. P. H. J. (2017). *Tools of the crook-infection strategies of fungal plant pathogens*. https://doi.org/10.1111/tpj.13810

Roesch, L. F. W., Fulthorpe, R. R., Riva, A., Casella, G., Hadwin, A. K. M., Kent, A. D., Daroub, S. H., Camargo, F. A. O., Farmerie, W. G., & Triplett, E. W. (2007). Pyrosequencing enumerates and contrasts soil microbial diversity. *ISME Journal, 1*(4), 283–290. https://doi.org/10.1038/ismej.2007.53

Rogosa, M., Krichevsky, M. I., & Colwell, R. R. (1986). *Coding microbiological data for computers*. New York: Springer. https://doi.org/10.1007/978-1-4612-4986-3

Rojas-Padilla, J., Chaparro-Encinas, L. A., Robles-Montoya, R. I., & de los Santos-Villalobos, S. (2020). Promoción de crecimiento en trigo (*Triticum turgidum* L. subsp. durum) por la co-inoculación de cepas nativas de *Bacillus aisladas* del Valle del Yaqui, México. *Nova Scientia, 12*(24), 1–27. https://doi.org/10.21640/ns.v12i24.2136

Rojas-Padilla, J., de-Bashan, L. E., Parra-Cota, F. I., Rocha-Estrada, J., & de los Santos-Villalobos, S. (2022). Microencapsulation of *Bacillus strains* for improving wheat (*Triticum turgidum* subsp. durum) growth and development. *Plants, 11*(21), 2920. https://doi.org/10.3390/plants11212920

Romano, I., Ventorino, V., & Pepe, O. (2020). Effectiveness of plant beneficial microbes: Overview of the methodological approaches for the assessment of root colonization and persistence. *Frontiers in Plant Science, 11*, 6. https://doi.org/10.3389/fpls.2020.00006

Romera, F. J., García, M. J., Lucena, C., Martínez-Medina, A., Aparicio, M. A., Ramos, J., Alcántara, E., Angulo, M., & Pérez-Vicente, R. (2019). Induced systemic resistance (ISR) and fe deficiency responses in dicot plants. *Frontiers in Plant Science, 10*(March), 1–17. https://doi.org/10.3389/fpls.2019.00287

Rosselló-Móra, R., & Amann, R. (2015). Past and future species definitions for Bacteria and Archaea. *Systematic and Applied Microbiology, 38*(4), 209–216. https://doi.org/10.1016/j.syapm.2015.02.001

Rotem, O., Pasternak, Z., & Jurkevitch, E. (2014). Bdellovibrio and like organisms. In E. Rosenberg, E. F. DeLong, S. Lory, E. Stackebrandt, & F. Thompson (Eds.), *The prokaryotes* (4th ed., pp. 4–17). Springer http://link.springer.com/referencework/10.1007/0-387-30742-7/page/2.

Ruan, Z., Ma, Q., & Sternfeld, E. (2020). Biofertilizers in China A potential strategy for China's sustainable agriculture current status and further perspectives. In *Sino-German agricultural centre, 2nd phase study.*

Rusnac, D. V., & Zheng, N. (2018). Overview of protein degradation in plant hormone signaling. In J. Hejátko, & T. Hakoshima (Eds.), *Plant structural biology: Hormonal regulations* (pp. 11–30). Springer International Publishing. https://doi.org/10.1007/978-3-319-91352-0_2

Ryu, C., Farag, M. A., Hu, C., Reddy, M. S., Kloepper, J. W., & Pare, P. W. (2004). Bacterial volatiles induced resistance in Arabidobsis. *Plant Physiology, 134*(March), 1017–1026. https://doi.org/10.1104/pp.103.026583.with

Sá, C., Cardoso, P., & Figueira, E. (2019). Alginate as a feature of osmotolerance differentiation among soil bacteria isolated from wild legumes growing in Portugal. *Science of the Total Environment, 681*, 312–319. https://doi.org/10.1016/j.scitotenv.2019.05.050

Sahu, P. K., & Brahmaprakash, G. P. (2016). Formulations of biofertilizers – approaches and advances. In D. P. Singh, H. B. Singh, & R. Prabha (Eds.), *Microbial inoculants in sustainable agricultural productivity: Vol. 2: Functional applications* (pp. 179–198). https://doi.org/10.1007/978-81-322-2644-4

Salazar-Cerezo, S., Martínez-Montiel, N., García-Sánchez, J., Pérez-y-Terrón, R., & Martínez-Contreras, R. D. (2018). Gibberellin biosynthesis and metabolism: A convergent route for plants, fungi and bacteria. *Microbiological Research, 208*(January), 85–98. https://doi.org/10.1016/j.micres.2018.01.010

de Salamone, I. E. G., Di Salvo, L. P., Ortega, J. S. E., Sorte, P. M. F. B., Urquiaga, S., & Teixeira, K. R. S. (2010). Field response of rice paddy crop to *Azospirillum inoculation*: Physiology of rhizosphere bacterial communities and the genetic diversity of endophytic bacteria in different parts of the plants. *Plant and Soil, 336*(1), 351–362. https://doi.org/10.1007/s11104-010-0487-y

Sammauria, R., Kumawat, S., Kumawat, P., Singh, J., & Jatwa, T. K. (2020). Microbial inoculants: Potential tool for sustainability of agricultural production systems. *Archives of Microbiology, 202*(4), 677–693. https://doi.org/10.1007/s00203-019-01795-w

Sang, M. K., & Kim, K. D. (2012). Plant growth-promoting rhizobacteria suppressive to *Phytophthora blight* affect microbial activities and communities in the rhizosphere of pepper (*Capsicum annuum* L.) in the field. *Applied Soil Ecology, 62*, 88–97. https://doi.org/10.1016/j.apsoil.2012.08.001

de los Santos-Villalobos, S., Díaz-Rodríguez, A. M., Ávila-Mascareño, M. F., Martínez-Vidales, A. D., & Parra-Cota, F. I. (2021). Colmena: A culture collection of native microorganisms for harnessing the agro-biotechnological potential in soils and contributing to food security. *Diversity, 13*(8). https://doi.org/10.3390/d13080337

Santos, M. S., Rodrigues, T. F., Nogueira, M. A., & Hungria, M. (2021). The challenge of combining high yields with environmentally friendly bioproducts: A review on the compatibility of pesticides with microbial inoculants. *Agronomy, 11*(5). https://doi.org/10.3390/agronomy11050870

de los Santos-Villalobos, S., Barrera-Galicia, G. C., Miranda-Salcedo, M. A., & Peña-Cabriales, J. J. (2012). Burkholderia cepacia XXVI siderophore with biocontrol capacity against *Colletotrichum gloeosporioides. World Journal of Microbiology and Biotechnology, 28*, 2615–2623.

de los Santos-Villalobos, S., Barrera-Galicia, G. C., Miranda-Salcedo, M. A., & Peña-Cabriales, J. J. (2012). Burkholderia cepacia XXVI siderophore with biocontrol capacity against *Colletotrichum gloeosporioides*. *World Journal of Microbiology and Biotechnology, 28*(8), 2615–2623. https://doi.org/10.1007/s11274-012-1071-9

dos Santos-Lopes, M. J., Dias-Filho, M. B., & Gurgel, E. S. C. (2021). Successful plant growth-promoting microbes: Inoculation methods and abiotic factors. *Frontiers in Sustainable Food Systems, 5*, 606454. https://doi.org/10.3389/fsufs.2021.606454

Santoyo, G., Sánchez-Yáñez, J. M., & de los Santos-Villalobos, S. (2019). Methods for detecting biocontrol and plant growth-promoting traits in rhizobacteria. In D. Reinhardt, & A. K. Sharma (Eds.), *Methods in rhizosphere biology research* (pp. 133–149). Springer Nature Singapore. https://doi.org/10.1007/978-981-13-5767-1_8

Saravanakumar, D., Lavanya, N., Muthumeena, K., Raguchander, T., & Samiyappan, R. (2009). Fluorescent pseudomonad mixtures mediate disease resistance in rice plants against sheath rot (*Sarocladium oryzae*) disease. *BioControl, 54*, 273–286. https://doi.org/10.1007/s10526-008-9166-9

Saravanakumar, K., Li, Y., Yu, C., Wang, Q. Q., Wang, M., Sun, J., Gao, J. X., & Chen, J. (2017). Effect of *Trichoderma harzianum* on maize rhizosphere microbiome and biocontrol of *Fusarium Stalk* rot. *Scientific Reports, 7*(1), 1–13. https://doi.org/10.1038/s41598-017-01680-w

Sartorius, A. G. (2022). *Lab Products and Bioprocess Solutions*. https://www.sartorius.com/en/products.

Schildkraut, C. L., Marmur, J., & Doty, P. (1961). The formation of hybrid DNA molecules and their use in studies of DNA homologies. *Journal of Molecular Biology, 3*(5), 595–617. https://doi.org/10.1016/S0022-2836(61)80024-7

Schlaeppi, K., & Mauch, F. (2010). Indolic secondary metabolites protect Arabidopsis from the oomycete pathogen *Phytophthora brassicae. Plant Signaling and Behavior, 5*(9), 1099–1101. https://doi.org/10.4161/psb.5.9.12410

Schlaeppi, K., Dombrowski, N., Oter, R. G., Ver Loren Van Themaat, E., & Schulze-Lefert, P. (2014). Quantitative divergence of the bacterial root microbiota in *Arabidopsis thaliana* relatives. *Proceedings of the National Academy of Sciences of the United States of America, 111*(2), 585–592. https://doi.org/10.1073/pnas.1321597111

Schünemann, R., Knaak, N., & Fiuza, L. M. (2014). Mode of action and specificity of *Bacillus thuringiensis* toxins in the control of caterpillars and stink bugs in soybean culture. *ISRN Microbiology*, 1–12. https://doi.org/10.1155/2014/135675, 2014.

Schwartz, S., Kent, W. J., Smit, A., Zhang, Z., Baertsch, R., Hardison, R. C., Haussler, D., & Miller, W. (2003). Human-mouse alignments with BLASTZ. *Genome Research, 13*(1), 103–107. https://doi.org/10.1101/gr.809403

Schwessinger, B., & Ronald, P. C. (2012). Plant innate immunity: Perception of conserved microbial signatures. *Annual Review of Plant Biology, 63*(1), 451–482. https://doi.org/10.1146/annurev-arplant-042811-105518

Seemann, T. (2014). Prokka: Rapid prokaryotic genome annotation. *Bioinformatics, 30*(14), 2068–2069. https://doi.org/10.1093/bioinformatics/btu153

Sentausa, E., & Fournier, P. E. (2013). Advantages and limitations of genomics in prokaryotic taxonomy. *Clinical Microbiology and Infection, 19*(9), 790–795. https://doi.org/10.1111/1469-0691.12181

Sessitsch, A., Pfaffenbichler, N., & Mitter, B. (2019). Microbiome applications from lab to field: Facing complexity. *Trends in Plant Science, 24*(3), 194–198. https://doi.org/10.1016/j.tplants.2018.12.004

Shameer, S., & Prasad, T. N. V. K. V. (2018). Plant growth promoting rhizobacteria for sustainable agricultural practices with special reference to biotic and abiotic stresses. *Plant Growth Regulation, 84*(3), 603–615. https://doi.org/10.1007/s10725-017-0365-1

Sharma, A., & Shouche, Y. (2014). Microbial culture collection (MCC) and international depositary authority (IDA) at national centre for cell science, pune. *Indian Journal of Microbiology, 54*(2), 129–133. https://doi.org/10.1007/s12088-014-0447-y

Sharma, S. B., Sayyed, R. Z., Trivedi, M. H., & Gobi, T. A. (2013). Phosphate solubilizing microbes: Sustainable approach for managing phosphorus deficiency in agricultural soils. *SpringerPlus, 2*(1), 1–14. https://doi.org/10.1186/2193-1801-2-587

Sharma, S. K., Kumar, R., Vaishnav, A., Sharma, P. K., Singh, U. B., & Sharma, A. K. (2017). Microbial cultures: Maintenance, preservation and registration. In A. Varma, & A. K. Sharma (Eds.), *Modern Tools and Techniques to understand microbes* (pp. 335–367). Springer Cham. https://doi.org/10.1007/978-3-319-49197-4

Sharma, V., Salwan, R., Sharma, P. N., & Gulati, A. (2017). Integrated translatome and proteome: Approach for accurate portraying of widespread multifunctional aspects of trichoderma. *Frontiers in Microbiology, 8*(August), 1–13. https://doi.org/10.3389/fmicb.2017.01602

Sharma, S. K., Saini, S., Verma, A., Sharma, P. K., Lal, R., Roy, M., Singh, U. B., Saxena, A. K., & Sharma, A. K. (2019). National agriculturally important microbial culture collection in the global context of microbial culture collection centres. *Proceedings of the National Academy of Sciences India Section B - Biological Sciences, 89*(2), 405–418. https://doi.org/10.1007/s40011-017-0882-8

Shen, W., Gao, N., Min, J., Shi, W., He, X., & Lin, X. (2016). Influences of past application rates of nitrogen and a catch crop on soil microbial communities between an intensive rotation. *Acta Agriculturae Scandinavica Section B: Soil and Plant Science, 66*(2), 97–106. https://doi.org/10.1080/09064710.2015.1072234

Silva, M. S., Arraes, F. B. M., Campos, M. de A., Grossi-de-Sa, M., Fernandez, D., Cândido, E. de S., Cardoso, M. H., Franco, O. L., & Grossi-de-Sa, M. F. (2018). Review: Potential biotechnological assets related to plant immunity modulation applicable in engineering disease-resistant crops. *Plant Science, 270*(February), 72–84. https://doi.org/10.1016/j.plantsci.2018.02.013

da Silva, M. F., de Souza Antônio, C., de Oliveira, P. J., Xavier, G. R., Rumjanek, N. G., Soares, L. H. de B., & Reis, V. M. (2012). Survival of endophytic bacteria in polymer-based inoculants and efficiency of their application to sugarcane survival of endophytic bacteria in polymer-based inoculants and efficiency of their application to sugarcane. *Plant Soil, 356*, 231–243. https://doi.org/10.1007/s11104-012-1242-3

Silveira, A. P. D. da, Sala, V. M. R., Cardoso, E. J. B. N., Labanca, E. G., & Cipriano, M. A. P. (2016). Nitrogen metabolism and growth of wheat plant under diazotrophic endophytic bacteria inoculation. *Applied Soil Ecology, 107*, 313–319. https://doi.org/10.1016/j.apsoil.2016.07.005

Simão, F. A., Waterhouse, R. M., Ioannidis, P., Kriventseva, E. V., & Zdobnov, E. M. (2015). BUSCO: Assessing genome assembly and annotation completeness with single-copy orthologs. *Bioinformatics, 31*(19), 3210–3212. https://doi.org/10.1093/bioinformatics/btv351

Singh, D. P., Singh, H. B., & Prabha, R. (2017). Plant-Microbe interactions in agro-ecological perspectives. In *Plant-microbe interactions in agro-ecological perspectives* (Vol. 2). https://doi.org/10.1007/978-981-10-6593-4

Singh, A., Shukla, N., Kabadwal, B. C., Tewari, A. K., & Kumar, J. (2018). Review on plant-trichoderma-pathogen interaction. *International Journal of Current Microbiology and Applied Sciences, 7*(2), 2382–2397. https://doi.org/10.20546/ijcmas.2018.702.291

Singh, S., Kumar, V., Dhanjal, D. S., & Singh, J. (2020). Biological control agents: Diversity, ecological significances, and biotechnological applications. In J. Singh, & A. N. Yadav (Eds.), *Natural bioactive products in sustainable agriculture* (pp. 31–44). Springer.

Smith, D., McCluskey, K., & Stackebrandt, E. (2014). Investment into the future of microbial resources: Culture collection funding models and BRC business plans for Biological Resource Centres. *SpringerPlus, 3*(1), 1–12. https://doi.org/10.1186/2193-1801-3-81

Soumare, A., Diedhiou, A. G., Thuita, M., & Hafidi, M. (2020). Exploiting biological nitrogen fixation: A route. *Plants*, 1–22.

Spaepen, S., & Vanderleyden, J. (2011). *Auxin and plant-microbe interactions.*

Spoel, S. H., Koornneef, A., Claessens, S. M. C., Korzelius, J. P., Van Pelt, J. A., Mueller, M. J., Buchala, A. J., Métraux, J. P., Brown, R., Kazan, K., Van Loon, L. C., Dong, X., & Pieterse, C. M. J. (2003). NPR1 modulates cross-talk between salicylate- and jasmonate-dependent defense pathways through a novel function in the cytosol. *Plant Cell, 15*(3), 760–770. https://doi.org/10.1105/tpc.009159

Srinivasan, K., & Mathivanan, N. (2009). Biological control of sunflower necrosis virus disease with powder and liquid formulations of plant growth promoting microbial consortia under field conditions. *Biological Control, 51*(3), 395–402. https://doi.org/10.1016/j.biocontrol.2009.07.013

Srivastava, A. K., & Gupta, S. (2011). Fed-Batch fermentation - design strategies. In*Comprehensive biotechnology, second edition* (2nd ed., Vol. 2). Elsevier B.V. https://doi.org/10.1016/B978-0-08-088504-9.00112-4

Staines, J., McGowan, V., & Skerman, V. B. D. (1986). *World directory of collections of cultures of microorganisms* (3d ed.). World Data Center.

Stamenković, S., Karabegović, I., Lazić, M., Nikolić, N., & Beškoski, V. (2018). Microbial fertilizers: A comprehensive review of current findings and future perspectives. *Spanish Journal of Agricultural Research, 16*(1), e09R01. https://doi.org/10.5424/sjar/2018161-12117

Staňková, H., Hastie, A. R., Chan, S., Vrána, J., Tulpová, Z., Kubaláková, M., Visendi, P., Hayashi, S., Luo, M., Batley, J., Edwards, D., Doležel, J., & Šimková, H. (2016). BioNano genome mapping of individual chromosomes supports physical mapping and sequence assembly in complex plant genomes. *Plant Biotechnology Journal, 14*(7), 1523–1531. https://doi.org/10.1111/pbi.12513

Stein, E., Molitor, A., Kogel, K. H., & Waller, F. (2008). Systemic resistance in Arabidopsis conferred by the mycorrhizal fungus *Piriformospora indica* requires jasmonic acid signaling and the cytoplasmic function of NPR1. *Plant and Cell Physiology, 49*(11), 1747–1751. https://doi.org/10.1093/pcp/pcn147

Stets, M. I., Campbell Alqueres, S. M., Maltempi Souza, E., de Oliveira Pedrosa, F., Schmid, M., Hartmann, A., & Magalhães Cruz, L. (2015). Quantification of *Azospirillum brasilense* FP2 bacteria in wheat roots by strain-specific quantitative PCR. *Applied and Environmental Microbiology, 81*(19), 6700–6709. https://doi.org/10.1128/AEM.01351-15

Strauß, T., Van Poecke, R. M. P., Strauß, A., Römer, P., Minsavage, G. V., Singh, S., Wolf, C., Strauß, A., Kim, S., Lee, H. A., Yeom, S. I., Parniske, M., Stall, R. E., Jones, J. B., Choi, D., Prins, M., & Lahaye, T. (2012). RNA-seq pinpoints a xanthomonas TAL-effector activated resistance gene in a large-crop genome. *Proceedings of the National Academy of Sciences of the United States of America, 109*(47), 19480–19485. https://doi.org/10.1073/pnas.1212415109

Strehmel, N., Böttcher, C., Schmidt, S., & Scheel, D. (2014). Profiling of secondary metabolites in root exudates of *Arabidopsis thaliana*. *Phytochemistry, 108*, 35–46. https://doi.org/10.1016/j.phytochem.2014.10.003

Stringlis, I. A., Yu, K., Feussner, K., de Jonge, R., Van Bentum, S., Van Verk, M. C., Berendsen, R. L., Bakker, P. A. H. M., Feussner, I., & Pieterse, C. M. (2018). MYB72-dependent coumarin exudation shapes root microbiome assembly to promote plant health. *Proceedings of the National Academy of Sciences, 115*(22), E5213–E5222.

Suh, S., Jiarong, P., & Toan, P. Van (2006). Quality control of biofertilizers. In *Biofertilizer manual. Forum for nuclear cooperation in asia (FNCA)* (pp. 112–124). Japan Atomic Industrial Forum (JAIF).

Sumi, C. D., Yang, B. W., Yeo, I. C., & Hahm, Y. T. (2015). Antimicrobial peptides of the genus Bacillus: A new era for antibiotics. *Canadian Journal of Microbiology, 61*(2), 93–103. https://doi.org/10.1139/cjm-2014-0613

Sun, H., Jiang, S., Jiang, C., Wu, C., Gao, M., & Wang, Q. (2021). A review of root exudates and rhizosphere microbiome for crop production. *Environmental Science and Pollution Research, 28*(39), 54497–54510. https://doi.org/10.1007/s11356-021-15838-7

Suthar, H., Hingurao, K., Vaghashiya, J., & Parmar, J. (2017). *Fermentation: A process for biofertilizer production* (pp. 229–252). https://doi.org/10.1007/978-981-10-6241-4_12

Tanaka, A., & Tsuge, T. (2000). Structural and functional complexity of the genomic region controlling AK-toxin biosynthesis and pathogenicity in the Japanese pear pathotype of *Alternaria alternata*. *Molecular Plant-Microbe Interactions, 13*(9), 975–986. https://doi.org/10.1094/MPMI.2000.13.9.975

Tanaka, A., Shiotani, H., Yamamoto, M., & Tsuge, T. (1999). Insertional mutagenesis and cloning of the genes required for biosynthesis of the host-specific AK-toxin in the Japanese pear pathotype of *Alternaria alternata*. *Molecular Plant-Microbe Interactions, 12*(8), 691–702. https://doi.org/10.1094/MPMI.1999.12.8.691

Tanaka, S., Brefort, T., Neidig, N., Djamei, A., Kahnt, J., Vermerris, W., Koenig, S., Feussner, K., Feussner, I., & Kahmann, R. (2014). A secreted ustilago maydis effector promotes virulence by targeting anthocyanin biosynthesis in maize. *ELife, 3*, 1–27. https://doi.org/10.7554/elife.01355

Tanizawa, Y., Fujisawa, T., & Nakamura, Y. (2018). Dfast: A flexible prokaryotic genome annotation pipeline for faster genome publication. *Bioinformatics, 34*(6), 1037–1039. https://doi.org/10.1093/bioinformatics/btx713

Tewari, S., & Arora, N. K. (2018). Role of salicylic acid from *Pseudomonas aeruginosa* PF23EPS+ in growth promotion of sunflower in saline soils infested with phytopathogen *Macrophomina phaseolina*. *Environmental Sustainability, 1*(1), 49–59. https://doi.org/10.1007/s42398-018-0002-6

Thakur, R., Verma, S., Gupta, S., Negi, G., & Bhardwaj, P. (2021). Role of soil health in plant disease management: A review. *Agricultural Reviews, 43*(Of), 70–76. https://doi.org/10.18805/ag.r-1856

Thi Nguyen, H. Y., & Tran, G. B. (2018). Optimization of fermentation conditions and media for production of glucose isomerase from *Bacillus megaterium* using response surface methodology. *Scientifica*. . https://doi.org/10.1155/2018/6842843, 2018.

Thijs, S., Weyens, N., Gkorezis, P., & Vangronsveld, J. (2016). Plant-Endophyte partnerships to assist petroleum hydrocarbon remediation S. In R. Steffan (Ed.), *Consequences of microbial interactions with Hydrocarbons, Oils, and lipids: Biodegradation and bioremediation* (pp. 1–34). Springer. https://doi.org/10.1007/978-3-319-44535-9

Thilagar, G., Bagyaraj, D. J., & Raoca, M. S. (2016). Selected microbial consortia developed for chilly reduces application of chemical fertilizers by 50% under field conditions. *Scientia Horticulturae, 198*, 27–35. https://doi.org/10.1016/j.scienta.2015.11.021

Thilakarathna, M. S., & Raizada, M. N. (2017). A meta-analysis of the effectiveness of diverse rhizobia inoculants on soybean traits under field conditions. *Soil Biology and Biochemistry, 105*, 177–196. https://doi.org/10.1016/j.soilbio.2016.11.022

Thomas, L., & Singh, I. (2019). Microbial biofertilizers: Types and applications. In B. Giri, R. Prasad, Q. S. Wu, & A. Varma (Eds.), *Biofertilizers for sustainable agriculture and environment. Soil biology* (Vol. 55, pp. 1–19). Springer.

Thomloudi, E. E., Tsalgatidou, P. C., Douka, D., Spantidos, T. N., Dimou, M., Venieraki, A., & Katinakis, P. (2019). Multistrain versus single-strain plant growth promoting microbial inoculants-the compatibility issue. *Hellenic Plant Protection Journal, 12*(2), 61–77. https://doi.org/10.2478/hppj-2019-0007

Thrash, A., Hoffmann, F., & Perkins, A. (2020). Toward a more holistic method of genome assembly assessment. *BMC Bioinformatics, 21*(4), 1–8. https://doi.org/10.1186/s12859-020-3382-4

Tian, B., Yang, J., & Zhang, K. Q. (2007). Bacteria used in the biological control of plant-parasitic nematodes: Populations, mechanisms of action, and future prospects. *FEMS Microbiology Ecology, 61*(2), 197–213. https://doi.org/10.1111/j.1574-6941.2007.00349.x

Tian, J., Ge, F., Zhang, D., Deng, S., & Liu, X. (2021). Roles of phosphate solubilizing microorganisms from managing soil phosphorus deficiency to mediating biogeochemical p cycle. *Biology, 10*(2), 1–19. https://doi.org/10.3390/biology10020158

Timmusk, S., Behers, L., Muthoni, J., Muraya, A., & Aronsson, A. C. (2017). Perspectives and challenges of microbial application for crop improvement. *Frontiers in Plant Science, 8*(February), 1–10. https://doi.org/10.3389/fpls.2017.00049

Tindall, B. J., Rosselló-Móra, R., Busse, H. J., Ludwig, W., & Kämpfer, P. (2010). Notes on the characterization of prokaryote strains for taxonomic purposes. *International Journal of Systematic and Evolutionary Microbiology, 60*(1), 249–266. https://doi.org/10.1099/ijs.0.016949-0

Trabelsi, D., & Mhamdi, R. (2013). Microbial inoculants and their impact on soil microbial communities: A review. *BioMed Research International.* https://doi.org/10.1155/2013/863240, 2013.

Trejo, A., Luz, E., Hartmann, A., Hernandez, J., Rothballer, M., Schmid, M., & Bashan, Y. (2012). Recycling waste debris of immobilized microalgae and plant growth-promoting bacteria from wastewater treatment as a resource to improve fertility of eroded desert soil. *Environmental and Experimental Botany, 75*, 65–73. https://doi.org/10.1016/j.envexpbot.2011.08.007

Tronsmo, A. M., Collinge, D. B., Djurle, A., Munk, L., Yuen, J., & Tronsmo, A. (2020). *Plant Pathology and Plant Diseases*. Wallingford, UK: CABI. https://doi.org/10.1079/9781789243185.0000

Tsalgatidou, P. C., Thomloudi, E. E., Baira, E., Papadimitriou, K., Skagia, A., Venieraki, A., & Katinakis, P. (2022). Integrated genomic and metabolomic analysis illuminates key secreted metabolites produced by the novel endophyte Bacillus halotolerans Cal.l.30 involved in diverse biological control activities. *Microorganisms, 10*(2). https://doi.org/10.3390/microorganisms10020399

Tsuge, T., Harimoto, Y., Akimitsu, K., Ohtani, K., Kodama, M., Akagi, Y., Egusa, M., Yamamoto, M., & Otani, H. (2013). Host-selective toxins produced by the plant pathogenic fungus *Alternaria alternata. FEMS Microbiology Reviews, 37*(1), 44–66. https://doi.org/10.1111/j.1574-6976.2012.00350.x

USDA. (2022). *Fertilizer Prices Spike in Leading U.S. Market in Late 2021, Just Ahead of 2022 Planting Season.* HYPERLINK http://www.ers.usda.gov/data-products/chart-gallery/gallery/chart-detail/?chartId=103194#:~:text=Fertilizer%20constitutes%20an%20average%20of,product%2C%20published%20in%20October%202021 www.ers.usda.gov/data-products/chart-gallery/gallery/chart-detail/?chartId=103194#:~:text=Fertilizer%20constitutes%20an%20average%20of,product%2C%20published%20in%20October%202021.

Valenzuela-Aragon, B., Parra-Cota, F. I., Santoyo, G., Arellano-Wattenbarger, G. L., & de los Santos-Villalobos, S. (2019). Plant-assisted selection: A promising alternative for in vivo identification of wheat (*Triticum turgidum* L. subsp. durum) growth promoting bacteria. *Plant and Soil, 435*(1–2), 367–384. https://doi.org/10.1007/s11104-018-03901-1

Valenzuela-Ruiz, V., Gálvez-Gamboa, G. T., Villa-Rodríguez, E. D., Parra-Cota, F. I., Santoyo, G., & de los Santos-Villalobos, S. (2020). Lipopéptidos producidos por agentes de control biológico del género Bacillus: Revisión de herramientas analíticas utilizadas para su estudio. *Revista Mexicana de Ciencias Agrícolas, 11*(2), 419–432. https://doi.org/10.29312/remexca.v11i2.2191

Valenzuela-Ruiz, V., Ayala-Zepeda, M., Arellano-Wattenbarger, G. L., Parra-Cota, F. I., García-Pereyra, G., Aviña-Martínez, G. N., & de los Santos-Villalobos, S. (2018). Las colecciones microbianas y su potencial contribución a la seguridad alimentaria actual y futura. *Revista Latinoamericana de Recursos Naturales, 14*(1), 18–25.

Valenzuela-Ruiz, V., Robles-Montoya, R. I., Parra-Cota, F. I., Santoyo, G., del Carmen Orozco-Mosqueda, M., Rodríguez-Ramírez, R., & de los Santos-Villalobos, S. (2019). Draft genome sequence of Bacillus paralicheniformis TRQ65, a biological control agent and plant growth-promoting bacterium isolated from wheat (*Triticum turgidum* subsp. durum) rhizosphere in the Yaqui Valley, Mexico. *3 Biotech, 9*(11), 1–7. https://doi.org/10.1007/s13205-019-1972-5

Valenzuela-Ruiz, V., Chávez-Luzania, R. A., Parra-Cota, F. I., Santoyo, G., & de los Santos-Villalobos, S. (2021). Extracellular polymeric substances from agriculturally important microorganisms. In A. Vaishnav, & D. K. Choudhary (Eds.), *Microbial polymers. Applications and ecological perspectives* (pp. 217–234). Springer. https://doi.org/10.1007/978-981-16-0045-6

Van Baarlen, P., Staats, M., & Van Kan, J. A. L. (2004). Induction of programmed cell death in lily by the fungal pathogen *Botrytis elliptica. Molecular Plant Pathology, 5*(6), 559–574. https://doi.org/10.1111/J.1364-3703.2004.00253.X

Vandamme, P., Pot, B., Gillis, M., Kersters, K., & Swings, J. (1996). Polyphasic taxonomy, a consensus approach to bacterial systematics. *Microbial Reviews, 2*(60), 407–438. https://doi.org/10.1007/s12088-007-0022-x

Vandana, U. K., Rajkumari, J., Singha, L. P., Satish, L., Alavilli, H., Sudheer, P. D. V. N., Chauhan, S., Ratnala, R., Satturu, V., Mazumder, P. B., & Pandey, P. (2021). The endophytic microbiome as a hotspot of synergistic interactions, with prospects of plant growth promotion. *Biology, 10*(2), 1–29. https://doi.org/10.3390/biology10020101

Van Loon, L. C. (2000). Helping plants to defend themselves: Biocontrol by disease-suppressing rhizobacteria. *Developments in Plant Genetics and Breeding, 6*(C), 203–213. https://doi.org/10.1016/S0168-7972(00)80123-1

Van Loon, L. C., & Bakker, P. A. H. M. (2005). Induced systemic resistance as a mechanism of disease suppression by rhizobacteria. In Z. A. Siddiqui (Ed.), *Pgpr: Biocontrol and biofertilization* (pp. 39–66). Springer. https://doi.org/10.1007/1-4020-4152-7_2

Varghese, N. J., Mukherjee, S., Ivanova, N., Konstantinidis, K. T., Mavrommatis, K., Kyrpides, N. C., & Pati, A. (2015). Microbial species delineation using whole genome sequences. *Nucleic Acids Research, 43*(14), 6761–6771. https://doi.org/10.1093/nar/gkv657

Vassilev, N., Vassileva, M., Lopez, A., Martos, V., Reyes, A., Maksimovic, I., Eichler-Löbermann, B., & Malusà, E. (2015). Unexploited potential of some biotechnological techniques for biofertilizer production and formulation. *Applied Microbiology and Biotechnology, 99*(12), 4983–4996. https://doi.org/10.1007/s00253-015-6656-4

Vassilev, N., Vassileva, M., Martos, V., Garcia del Moral, L. F., Kowalska, J., Tylkowski, B., & Malusá, E. (2020). Formulation of microbial inoculants by encapsulation in natural polysaccharides: Focus on beneficial properties of carrier additives and derivatives. *Frontiers in Plant Science, 11*(March), 1–9. https://doi.org/10.3389/fpls.2020.00270

Vejan, P., Khadiran, T., Abdullah, R., Ismail, S., & Dadrasnia, A. (2019). Encapsulation of plant growth promoting rhizobacteria—prospects and potential in agricultural sector: A review. *Journal of Plant Nutrition, 42*(19), 2600–2623. https://doi.org/10.1080/01904167.2019.1659330

Velásquez, A. C., Castroverde, C. D. M., & He, S. Y. (2018). Plant and pathogen warfare under changing climate conditions. *Current Biology, 28*(10), R619–R634. https://doi.org/10.1016/j.cub.2018.03.054

Verhagen, B. W. M., Glazebrook, J., Zhu, T., Chang, H. S., Van Loon, L. C., & Pieterse, C. M. J. (2004). The transcriptome of rhizobacteria-induced systemic resistance in *Arabidopsis*. *Molecular Plant-Microbe Interactions, 17*(8), 895–908. https://doi.org/10.1094/MPMI.2004.17.8.895

Verma, J. P., Jaiswal, D. K., & Maurya, P. K. (2016). Screening of bacterial strains for developing effective pesticide-tolerant plant growth-promoting microbial consortia from rhizosphere soils of vegetable fields of eastern Uttar Pradesh, India. *Energy, Ecology and Environment, 1*(6), 408–418. https://doi.org/10.1007/s40974-016-0028-5

Veselova, S. V., Nuzhnaya, T. V., & Maksimov, I. V. (2015). Role of jasmonic acid in interaction of plants with plant growth promoting rhizobacteria during fungal pathogenesis. *Jasmonic Acid: Biosynthesis, Functions and Role in Plant Development*, 33–66. October.

Vessey, J. K. (2003). Plant growth promoting rhizobacteria as biofertilizers. *Plant and Soil, 255*(2), 571–586. https://doi.org/10.1023/A:1026037216893

Viji, G., Uddin, W., & Romaine, C. P. (2003). Suppression of gray leaf spot (blast) of perennial ryegrass turf by *Pseudomonas aeruginosa* from spent mushroom substrate. *Biological Control, 26*(3), 233–243. https://doi.org/10.1016/S1049-9644(02)00170-6

Vikrant, K., Giri, B. S., Raza, N., Roy, K., Kim, K. H., Rai, B. N., & Singh, R. S. (2018). Recent advancements in bioremediation of dye: Current status and challenges. *Bioresource Technology, 253*, 355–367. https://doi.org/10.1016/j.biortech.2018.01.029

Villa-Rodríguez, E., Parra-Cota, F., Castro-Longoria, E., López-Cervantes, J., & de los Santos-Villalobos, S. (2019). *Bacillus* subtilis TE3: A promising biological control agent against bipolaris sorokiniana, the causal agent of spot blotch in wheat (*Triticum turgidum* L. subsp. durum). *Biological Control, 132*, 135–143. https://doi.org/10.1016/j.biocontrol.2019.02.012

Villa-Rodriguez, E., Moreno-Ulloa, A., Castro-Longoria, E., Parra-Cota, F. I., & de los Santos-Villalobos, S. (2021). Integrated omics approaches for deciphering antifungal metabolites produced by a novel *Bacillus* species, B. cabrialesii TE3T, against the spot blotch disease of wheat (*Triticum turgidum* L. subsp. durum). *Microbiological Research, 251*, 126826. https://doi.org/10.1016/j.micres.2021.126826

Villarreal-Delgado, M. F., Villa-Rodríguez, E. D., Cira-Chávez, L. A., Estrada-Alvarado, M. I., Parra-Cota, F. I., & De los Santos-Villalobos, S. (2018). El género Bacillus como agente de control biológico y sus implicaciones en la bioseguridad agrícola. *Mexican Journal of Phytopathology, 36*(1), 95–130. https://doi.org/10.18781/r.mex.fit.1706-5

Vorholt, J. A. (2012). Microbial life in the phyllosphere. *Nature Reviews Microbiology, 10*(12), 828–840. https://doi.org/10.1038/nrmicro2910

Vurukonda, S. S. K. P., Giovanardi, D., & Stefani, E. (2018). Plant growth promoting and biocontrol activity of streptomyces spp. as endophytes. *International Journal of Molecular Sciences, 19*(4). https://doi.org/10.3390/ijms19040952

Wahid, F., Fahad, S., Danish, S., Adnan, M., Yue, Z., Saud, S., Siddiqui, M. H., Brtnicky, M., Hammerschmiedt, T., & Datta, R. (2020). Sustainable management with mycorrhizae and phosphate solubilizing bacteria for enhanced phosphorus uptake in calcareous soils. *Agriculture (Switzerland), 10*(8), 1–14. https://doi.org/10.3390/agriculture10080334

Wahl, R., Wippel, K., Goos, S., Kämper, J., & Sauer, N. (2010). A novel high-affinity sucrose transporter is required for virulence of the plant pathogen *Ustilago maydis*. *PLoS Biology, 8*(2). https://doi.org/10.1371/journal.pbio.1000303

Wang, X., Wang, L., Wang, J., Jin, P., Liu, H., & Zheng, Y. (2014). Bacillus cereus AR156-induced resistance to colletotrichum acutatum is associated with priming of defense responses in loquat fruit. *PLoS ONE, 9*(11), 3–10. https://doi.org/10.1371/journal.pone.0112494

Wang, J., Li, Q., Xu, S., Zhao, W., Lei, Y., Song, C., & Huang, Z. (2018). Traits-based integration of multi-species inoculants facilitates shifts of indigenous soil bacterial community. *Frontiers in Microbiology, 9*(July), 1–13. https://doi.org/10.3389/fmicb.2018.01692

Wang, C., Zeng, Z. Q., & Zhuang, W. Y. (2021). Comparative molecular evolution of chitinases in ascomycota with emphasis on mycoparasitism lifestyle. *Microbial Genomics, 7*(9). https://doi.org/10.1099/MGEN.0.000646

Wang, Y., Li, X., Fan, B., Zhu, C., & Chen, Z. (2021). Regulation and function of defense-related callose deposition in plants. *International Journal of Molecular Sciences, 22*(5), 1–15. https://doi.org/10.3390/ijms22052393

Wang, Y., Xu, Z., Zhu, P., Liu, Y., Zhang, Z., Mastuda, Y., Toyoda, H., & Xu, L. (2010). Postharvest biological control of melon pathogens using *Bacillus subtilis* EXWB1. *Journal of Plant Pathology, 92*(3), 645–652.

Wani, P. A., Khan, M. S., & Zaidi, A. (2007). Effect of metal tolerant plant growth promoting bradyrhizobium sp. (vigna) on growth, symbiosis, seed yield and metal uptake by greengram plants. *Chemosphere, 70*(1), 36–45. https://doi.org/10.1016/j.chemosphere.2007.07.028

Wattam, A. R., Davis, J. J., Assaf, R., Boisvert, S., Brettin, T., Bun, C., Conrad, N., Dietrich, E. M., Disz, T., Gabbard, J. L., Gerdes, S., Henry, C. S., Kenyon, R. W., Machi, D., Mao, C., Nordberg, E. K., Olsen, G. J., Murphy-Olson, D. E., Olson, R., et al. (2017). Improvements to PATRIC, the all-bacterial bioinformatics database and analysis resource center. *Nucleic Acids Research, 45*(D1), D535–D542. https://doi.org/10.1093/nar/gkw1017

Wayne, L. G., Brenner, D. J., Colwell, R., Grimont, P. A. D., Kandler, O., Krichevsky, M. I., Moore, L. H., Moore, W. E. C., Murray, R. G. E., Stackebrandt, E., Starr, M. P., & Truper, H. G. (1987). Report of the ad hoc committee on reconciliation of approaches to bacterial systematics. *International Journal of Systematic Bacteriology, 37*(4), 463–464. https://doi.org/10.1099/00207713-37-4-463

WDCM. (2022). *Statistics*. http://www.wfcc.info/ccinfo/statistics/.

Wee, Y., Bhyan, S. B., Liu, Y., Lu, J., Li, X., & Zhao, M. (2019). The bioinformatics tools for the genome assembly and analysis based on third-generation sequencing. *Briefings in Functional Genomics, 18*(1), 1–12. https://doi.org/10.1093/bfgp/ely037

Weisburg, W. G., Barns, S. M., Pelletier, D. A., & Gene-Trak, D. J. L. (1991). 16S ribosomal DNA amplification for phylogenetic study. *Journal of Bacteriology, 173*(2), 697–703. https://doi.org/10.1186/s12903-020-01262-9

Whitman, W. B., Coleman, D. C., & Wiebe, W. J. (1998). Prokaryotes: The unseen majority. *Proceedings of the National Academy of Sciences of the United States of America, 95*, 6578–6583. https://doi.org/10.1038/s41437-017-0034-1

Wong, J. W. H., Plett, K. L., Natera, S. H. A., Roessner, U., Anderson, I. C., & Plett, J. M. (2020). Comparative metabolomics implicates threitol as a fungal signal supporting colonization of *Armillaria luteobubalina* on eucalypt roots. *Plant Cell and Environment, 43*(2), 374–386. https://doi.org/10.1111/pce.13672

Wongdee, J., Boonkerd, N., Teaumroong, N., Tittabutr, P., & Giraud, E. (2018). Regulation of nitrogen fixation in bradyrhizobium sp. strain DOA9 involves two distinct NifA regulatory proteins that are functionally redundant during symbiosis but not during free-living growth. *Frontiers in Microbiology, 9*(July), 1–11. https://doi.org/10.3389/fmicb.2018.01644

Wu, Z., Guo, L., Zhao, Y., & Li, C. (2014). Effect of free and encapsulated raoultella planticola rs-2 on cotton growth promotion under salt stress. *Journal of Plant Nutrition, 37*, 1187–1201. https://doi.org/10.1080/01904167.2014.881865

Wu, L., Sun, Q., Desmeth, P., Sugawara, H., Xu, Z., McCluskey, K., Smith, D., Alexander, V., Lima, N., Ohkuma, M., Robert, V., Zhou, Y., Li, J., Fan, G., Ingsriswang, S., Ozerskaya, S., & Ma, J. (2017). World data centre for microorganisms: An information infrastructure to explore and utilize preserved microbial strains worldwide. *Nucleic Acids Research, 45*(D1), D611–D618. https://doi.org/10.1093/nar/gkw903

Wubben, J. P., Mulder, W., Have, A. Ten, Van Kan, J. A. L., & Visser, J. (1999). Cloning and partial characterization of endopolygalacturonase genes from *Botrytis cinema. Applied and Environmental Microbiology, 65*(4), 1596–1602. https://doi.org/10.1128/aem.65.4.1596-1602.1999

Xi, W., Gao, Y., Cheng, Z., Chen, C., Han, M., Yang, P., Xiong, G., & Ning, K. (2019). Using QC-blind for quality control and contamination screening of bacteria DNA sequencing data without reference genome. *Frontiers in Microbiology, 10*(July), 1–15. https://doi.org/10.3389/fmicb.2019.01560

Xiao-Jing, xu, li-Qun, Z., You-Yong, Z., & Wen-Hua, T. (2005). Improving biocontrol effect of pseudomonas fluorescens P5 on plant diseases by genetic modification with chitinase gene. *Chinese Journal of Agricultural Biotechnology, 2*(1), 23–27. https://doi.org/10.1079/CJB200553

Xie, S., Yu, H., Li, E., Wang, Y., Liu, J., & Jiang, H. (2019). Identification of miRNAs involved in Bacillus velezensis FZB42-activated induced systemic resistance in maize. *International Journal of Molecular Sciences, 20*(20), 5057.

Xiong, W., Guo, S., Jousset, A., Zhao, Q., Wu, H., Li, R., Kowalchuk, G. A., & Shen, Q. (2017). Bio-fertilizer application induces soil suppressiveness against fusarium wilt disease by reshaping the soil microbiome. *Soil Biology and Biochemistry, 114*, 238–247. https://doi.org/10.1016/j.soilbio.2017.07.016

Xu, X. M., & Jeger, M. J. (2013). Theoretical modeling suggests that synergy may result from combined use of two biocontrol agents for controlling foliar pathogens under spatial heterogeneous conditions. *Phytopathology, 103*(8), 768–775. https://doi.org/10.1094/PHYTO-10-12-0266-R

Xu, X. M., Jeffries, P., Pautasso, M., & Jeger, M. J. (2011). Combined use of biocontrol agents to manage plant diseases in theory and practice. *Phytopathology, 101*(9), 1024–1031. https://doi.org/10.1094/PHYTO-08-10-0216

Xu, L., Naylor, D., Dong, Z., Simmons, T., Pierroz, G., Hixson, K. K., Kim, Y. M., Zink, E. M., Engbrecht, K. M., Wang, Y., Gao, C., DeGraaf, S., Madera, M. A., Sievert, J. A., Hollingsworth, J., Birdseye, D., Scheller, H. V., Hutmacher, R., Dahlberg, J., et al. (2018). Drought delays development of the sorghum root microbiome and enriches for monoderm bacteria. *Proceedings of the National Academy of Sciences of the United States of America, 115*(18), E4284–E4293. https://doi.org/10.1073/pnas.1717308115

Xu, Z., Wang, M., Du, J., Huang, T., Liu, J., Dong, T., & Chen, Y. (2020). Isolation of burkholderia sp. HQB-1, A promising biocontrol bacteria to protect banana against fusarium wilt through phenazine-1-carboxylic acid secretion. *Frontiers in Microbiology, 11*(December), 1–12. https://doi.org/10.3389/fmicb.2020.605152

Xu, S., Liu, Y. X., Cernava, T., Wang, H., Zhou, Y., Xia, T., Cao, S., Berg, G., Shen, X. X., Wen, Z., Li, C., Qu, B., Ruan, H., Chai, Y., Zhou, X., Ma, Z., Shi, Y., Yu, Y., Bai, Y., & Chen, Y. (2022). Fusarium fruiting body microbiome member *Pantoea agglomerans* inhibits fungal pathogenesis by targeting lipid rafts. *Nature Microbiology, 7*(6), 831–843. https://doi.org/10.1038/s41564-022-01131-x

Yadav, A. N. (2020). Plant Microbiomes for sustainable agriculture: Current research and future challenges. In A. Yadav, J. Singh, A. Rastegari, & N. Yadav (Eds.), *Plant microbiomes for sustainable agriculture. Sustainable development and biodiversity* (Vol 25). Cham: Springer. https://doi.org/10.1007/978-3-030-38453-1_16

Ye, H., Lu, C., & Lin, Q. (2019). Investigation of the spatial heterogeneity of soil microbial biomass carbon and nitrogen under long-term fertilizations in fluvo-aquic soil. *PLoS ONE, 14*(4), 1–21. https://doi.org/10.1371/journal.pone.0209635

Yoon, S. H., Ha, S. min, Lim, J., Kwon, S., & Chun, J. (2017). A large-scale evaluation of algorithms to calculate average nucleotide identity. *Antonie Van Leeuwenhoek, International Journal of General and Molecular Microbiology, 110*(10), 1281–1286. https://doi.org/10.1007/s10482-017-0844-4

Yu, X., Ai, C., Xin, L., & Zhou, G. (2011). The siderophore-producing bacterium, *Bacillus subtilis* CAS15, has a biocontrol effect on fusarium wilt and promotes the growth of pepper. *European Journal of Soil Biology, 47*(2), 138–145. https://doi.org/10.1016/j.ejsobi.2010.11.001

Yu, X., Li, Y., Zhang, C., Liu, H., Liu, J., Zheng, W., Kang, X., Leng, X., Zhao, K., Gu, Y., Zhang, X., Xiang, Q., & Chen, Q. (2014). Culturable heavy metal-resistant and plant growth promoting bacteria in V-Ti magnetite mine tailing soil from Panzhihua, China. *PLoS ONE, 9*(9). https://doi.org/10.1371/journal.pone.0106618

Yu, P., He, X., Baer, M., Beirinckx, S., Tian, T., Moya, Y. A. T., Zhang, X., Deichmann, M., Frey, F. P., Bresgen, V., Li, C., Razavi, B. S., Schaaf, G., Wirén, N. von, Su, Z., Bucher, M., Tsuda, K., Goormachtig, S., Chen, X., & Hochholdinger, F. (2021). Plant flavones enrich rhizosphere Oxalobacteraceae to improve maize performance under nitrogen deprivation. *Nature Plants, 7*, 481–499.

Zerbino, D. R., & Birney, E. (2008). Velvet: Algorithms for de novo short read assembly using de Bruijn graphs. *Genome Research, 18*(5), 821–829. https://doi.org/10.1101/gr.074492.107

Zhang, S., Ma, Y., Jiang, W., Meng, L., Cao, X., Hu, J., Chen, J., & Li, J. (2020). Development of a strain-specific quantification method for monitoring *Bacillus amyloliquefaciens* TF28 in the rhizospheric soil of soybean. *Molecular Biotechnology, 62*(10), 521–533. https://doi.org/10.1007/s12033-020-00268-6

Zhang, Y., Shan, Q., Wang, Y., Chen, K., Liang, Z., Li, J., Zhang, Y., Zhang, K., Liu, J., Voytas, D. F., Zheng, X., & Gao, C. (2013). Rapid and efficient gene modification in rice and brachypodium using TALENs. *Molecular Plant, 6*(4), 1365–1368. https://doi.org/10.1093/mp/sss162

Zhang, N., Yang, D., Wang, D., Miao, Y., Shao, J., Zhou, X., Xu, Z., Li, Q., Feng, H., Li, S., Shen, Q., & Zhang, R. (2015). Whole transcriptomic analysis of the plant-beneficial rhizobacterium *Bacillus amyloliquefaciens* SQR9 during enhanced biofilm formation regulated by maize root exudates. *BMC Genomics, 16*(1), 1–20. https://doi.org/10.1186/s12864-015-1825-5

Zhang, N., Yang, D., Kendall, J. R. A., Borriss, R., Druzhinina, I. S., Kubicek, C. P., Shen, Q., & Zhang, R. (2016). Comparative genomic analysis of Bacillus amyloliquefaciens and *Bacillus subtilis* reveals evolutional traits for adaptation to plant-associated habitats. *Frontiers in Microbiology, 7*(December). https://doi.org/10.3389/fmicb.2016.02039

Zhang, Q., Saleem, M., & Wang, C. (2017). Probiotic strain Stenotrophomonas acidaminiphila BJ1 degrades and reduces chlorothalonil toxicity to soil enzymes, microbial communities and plant roots. *AMB Express, 7*(1), 4–11. https://doi.org/10.1186/s13568-017-0530-y

Zhang, L., Yan, C., Guo, Q., Zhang, J., & Ruiz-Menjivar, J. (2018). The impact of agricultural chemical inputs on environment: Global evidence from informetrics analysis and visualization. *International Journal of Low-Carbon Technologies, 13*(4), 338–352. https://doi.org/10.1093/ijlct/cty039

Zhang, Z., Yuen, G. Y., Sarath, G., & Penheiter, A. R. (2001). Chitinases from the plant disease biocontrol agent, Stenotrophomonas maltophilia C3. *Phytopathology, 91*(2), 204–211. https://doi.org/10.1094/PHYTO.2001.91.2.204

Zhao, P., Li, P., Wu, S., Zhou, M., Zhi, R., & Gao, H. (2019). Volatile organic compounds (VOCs) from *Bacillus subtilis* CF-3 reduce anthracnose and elicit active defense responses in harvested litchi fruits. *AMB Express, 9*(1). https://doi.org/10.1186/s13568-019-0841-2

Zhao, Y., Yao, Y., Xu, H., Xie, Z., Guo, J., Qi, Z., & Jiang, H. (2022). Soil metabolomics and bacterial functional traits revealed the responses of rhizosphere soil bacterial community to long-term continuous cropping of Tibetan barley. *PeerJ, 10*. https://doi.org/10.7717/peerj.13254

Zhu, Y. G., Xiong, C., Wei, Z., Chen, Q. L., Ma, B., Zhou, S. Y. D., Tan, J., Zhang, L. M., Cui, H. L., & Duan, G. L. (2022). Impacts of global change on the phyllosphere microbiome. *New Phytologist, 234*(6), 1977–1986. https://doi.org/10.1111/nph.17928

Zimin, A. V., Marçais, G., Puiu, D., Roberts, M., Salzberg, S. L., & Yorke, J. A. (2013). The MaSuRCA genome assembler. *Bioinformatics, 29*(21), 2669–2677. https://doi.org/10.1093/bioinformatics/btt476

Zvinavashe, A. T., Mardad, I., Mhada, M., Kouisni, L., & Marelli, B. (2021). Engineering the plant microenvironment to facilitate plant-growth-promoting microbe association. *Journal of Agricultural and Food Chemistry, 69*(45), 13270–13285. https://doi.org/10.1021/acs.jafc.1c00138

Index

Note: 'Page numbers followed by "f" indicate figures and "t" indicate tables.'

J

L

M

Q

R

S

T

U

V

W

Printed in the United States
by Baker & Taylor Publisher Services